The
Screech
Owl
Companion

The
Screech
Owl.
Companion

Jim Wright and Scott Weston

Timber Press
Portland, Oregon

Published in 2024 by Timber Press, Inc.,
a subsidiary of Workman Publishing Co., Inc.,
a subsidiary of Hachette Book Group, Inc.
1290 Avenue of the Americas
New York, New York 10104
timberpress.com

Printed in China on responsibly sourced paper
Text and cover design by Leigh Thomas

The publisher is not responsible for websites (or their content) that are not owned by the publisher.

The Hachette Speakers Bureau provides a wide range of authors for speaking events. To find out more, go to hachettespeakersbureau.com or email HachetteSpeakers@hbgusa.com.

ISBN: 978-1-64326-189-8
Catalog records for this book are available from the Library of Congress and the British Library.

In memory of my father, John S. Wright, and my brother, John S. Wright Jr.

—Jim Wright

In memory of my grandfather, Edwin Pumala

—Scott Weston

To know the owls, class Aves, order Strigiformes, you don't need a degree from Cornell or a grant from Exxon. You needn't go on exotic safaris or buy a lot of expensive gadgets. Only pay heed to whatever district you live in and listen to the night surrounding you: There's more going on under your own window than you can absorb in a full and fruitful lifetime.

–Jonathan Maslow, *The Owl Papers*

Contents

Introduction

By Jim Wright

Four moments shaped this book.

1

Two decades ago, on a sunny spring morning, a bird-watching acquaintance mentioned that she'd seen a screech owl perched in the nest box on her property. When I expressed my awe, she shrugged and replied, "It's no big deal. I bet you'll get one if you put up an owl box."

2

The following autumn, my wife Patty and I installed an owl box with a gable roof about 50 feet from our living room window. We placed the nest box on a red maple in the adjacent nature preserve and glanced at it wistfully from time to time. No sign of activity.

This red-morph Eastern Screech Owl appeared magically one October afternoon.

One day, our friend the marsh warden put a handwritten note on our door: "Great news. You have a red-phase screech owl in your box!" We were so excited by the news that we saved the note and pinned it to a bulletin board. But no matter how often we looked, we saw nothing but the owl box's empty opening.

All that changed abruptly one Sunday afternoon when a most enchanting little owl poked its head out. Its arrival was so momentous that Jim still remembers the day: October 17, 2004.

Ever since childhood, I had heard that owls possessed magical powers. Could this little reddish-brown raptor be some sort of sign? If so, of what?

I should mention that Patty and I live in northern New Jersey, maybe 18 miles from Yankee Stadium as the raptor flies. On that Sunday, the Yankees were on the brink of sweeping the Red Sox in the American League Championship Series. In fact, the Bronx Bombers led Boston in the bottom of the ninth that night. Three outs away from victory.

Then the strangest thing happened. The Red Sox somehow tied the game and beat the Yankees in extra innings. The Red Sox proceeded to beat the Bombers three more times and then swept the St. Louis Cardinals in the World Series. From the moment the screech owl appeared in our yard that October afternoon, the Red Sox won every game.

If you believe in omens, our new arrival more than fit the bill. But since we lived in the Yankees' heartland, we told no one about our mysterious guest, lest it got blamed for the home team's collapse. But we knew magic was in the air, especially after dark.

3

In the spring of 2005, a grayish-brown screech owl appeared. No more Red Sox mojo, but we soon had something even better: a real-live owl for a neighbor. At dusk each day, the owl perched in the nest-box opening before setting off for parts unknown. We wondered what was happening inside that shoebox-size home, but we refrained from peering inside and interfering.

One afternoon in late May, I looked over at the nest box and saw a fuzzy little white head sticking out of the opening. My eyes nearly popped out of my head.

Soon after, my curiosity kicked in. A family of owls had been living near my house since early spring, and I knew nothing about them. What was going on in that little pine house? When did they have eggs? How many? When did the eggs hatch? And what would happen next? I started to learn as much about screech owls as I could.

4

The rains came down heavy in the spring of 2006, especially in late May when we saw an owlet at the opening each afternoon. Until we didn't.

After fretting for a couple of days, we crept toward the screech box, located about 8 feet up the maple, and found a dead, wet owlet on the ground. When we looked up at the box, we saw that the wood along the top of the gable roof had warped. The roof was no longer watertight.

We waited a few days. After seeing no more signs of life, we opened the box. Inside was nothing but a pile of wet woodchips and assorted bird feathers. What had happened, and why had we been powerless to help the owlet?

We got a shovel, buried the owlet, and said a little prayer. We realized then and there that we needed to know more about these mysterious nocturnal raptors. And we needed to install a camera inside the owl box before the next nesting season.

This book tells the story about the joys and heartaches of providing a nearby home for screech owls, the most amazing creatures you'll seldom see. It's also a guidebook on how you can participate in the adventure.

We offer one word of warning. Although inducing a pair of screech owls to start a family might be as easy as putting up an owl box in the right location, brace yourself for an emotional rollercoaster ride. This is especially true if you install a miniature video camera inside the box to watch during nesting season—the next natural step.

Sure, you'll rejoice at seeing your first screech perch in the opening of the owl box. But as soon as the owl disappears for a few days, you'll fret that it won't come back. And, sure, witnessing an owlet emerge from its thin white shell is unforgettable. But you'll also worry that those owlets, fragile as cotton balls, can't possibly survive. Or that mom has left her babies for far too long on a chilly or rainy night.

This book aims to make those ups and downs as enriching as possible. The ride will still be bumpy, with plenty of twists and turns, but my coauthor Scott Weston and I will help you reduce the avoidable missteps. One chapter showcases Scott's ingenious design for building a squirrel-resistant screech box.

We also include a primer on installing a camera to monitor the inside of the nest box. The cameras will enable you to watch and listen to your owl family during nesting season whenever you want. With the help of the right software and You-Tube, you can even rewind the video for up to 12 hours to see what you missed when you weren't tuned in and record the most memorable moments as separate videos.

These tiny digital cameras have changed the way people view these fascinating birds and have expanded our knowledge base. The cameras have confirmed that while all screeches tend to behave similarly, you should expect the unexpected. Screech owls are like people. No two are the same.

In researching this book, Scott and I picked the brains of several folks who livestream their screech-owl cams, from skilled first-timers to veterans who've used nestcams and other cameras for a decade. This book includes their hard-earned know-how—from advice on putting up an owl box to the benefits of installing a second box and camera.

Most of what we have learned came the hard way, through trial, error, and observation. We hope this book will spare you some of the headaches and heartaches and enrich your appreciation for these incredible creatures.

Screech owls have often been cloaked in mystery and superstition.

Rejoice!

I rejoice that there are owls. Let them do the idiotic and maniacal hooting for men. It is a sound admirably suited to swamps and twilight woods which no day illustrates, suggesting a vast and undeveloped nature which men have not recognized. They represent the stark twilight and unsatisfied thoughts which all have.

—Henry David Thoreau, *Walden*

Meet Your Neighbors

Screech Owl Basics

You heard an eerie call near the garden one night. You saw a mysterious little owl poke its head from an old woodpecker hole or perch on a nearby branch at dusk.

Ring a bell? Then you likely have a screech owl living in your neighborhood. These tiny raptors may weigh about as much as a hockey puck, but few birds rival their charisma, parenting skills, or hunting prowess.

Boise State University professor Jim Belthoff has studied many owl species since the early 1990s, but screech owls are his favorites. "They have these large yellow eyes," he says. "They have ear tufts. They give wonderful vocalizations. They have these really interesting plumage morphs. I don't know if I'm allowed to say that as a biologist, but, in my mind, they are the quintessential owl."

Now could be your chance to entice one to move in next door.

An Eastern Screech Owl peers out from a natural tree cavity.

Although screech owls increasingly like to live near humans, they live in a parallel universe. They are predators that hunt by night—no small feat. They go to sleep when we get up and awaken when we're settling in after a long day. Despite being relatively common in many areas in and around major cities, screech owls aren't a backyard bird that quickly comes to mind.

Also, forget Harry Potter and the Hogwarts instructions that "students may bring an owl or a cat or a toad." Screech owls aren't pets, even though some owl-box hosts probably can't help but wonder if a screech would make a good one. Our advice: Don't even consider it. In most places, owning a pet owl is illegal and—unless it is a rehab animal—cruel. Owls belong in the wild, not in captivity.

The Owls of Harry Potter

Although the Harry Potter books are credited for much of the public's recent fascination with owls, screech owls received short shrift in the novels for one simple reason. Harry Potter and Ron Weasley are British and screech owls live on the other side of the Atlantic. *Harry Potter and the Order of the Phoenix* mentions a screech, though it may well be a Barn Owl:

"But at that exact moment, a screech owl swooped in through the kitchen window. Narrowly missing the top of Uncle Vernon's head, it soared across the kitchen, dropped a large parchment envelope it was carrying in its beak at Harry's and my feet, and turned gracefully, the tip of its wings just brushing the top of the fridge, then flew outside again."

Ron Weasley's first pet owl was Pigwidgeon, a Eurasian Scops Owl—the Eastern Hemisphere equivalent of a screech owl. Also known as Pig, the owl was responsible for handling Ron's mail. In *Harry Potter and the Half-Blood Prince*, Ron bought Pig food at Eeylops Owl Emporium, purveyor of owls and owl supplies.

Eeylops Owl Emporium was a setting in the film version of *Harry Potter and the Philosopher's Stone.*

Owl researcher Fred Gehlbach wrote this classic text on screech owls.

A Classic Text

For decades, the bible for screech owl lovers has been *The Eastern Screech Owl* (1994), a textbook by the legendary Frederick R. Gehlbach, who spent several decades researching virtually every aspect of these owls' behavior, with all the facts and figures to prove it. The author clearly has a love for his subject matter. It begins with the acknowledgments, in which he thanks 659 Eastern Screech Owls "for enriching my life by revealing theirs, and for their resilience in our urbanizing world."

Although the book does lapse into some technical scientific jargon, screech owl lovers appreciate Gehlbach's voluminous research and enjoy the stray facts that pop up along the way. Here's a sample:

> "At first, I tacitly assumed that all Eastern Screech Owls were a common entity, a subtle form of human prejudice. But I quickly found that their individualities, including sizes and plumages, were as distinctive as the countenances and clothing of humans."

> "Female screech owls average 17 percent heavier than their mates, and the larger ones lay more, heavier eggs and lose less weight while nesting, hence fledge more offspring."

> "By St. Patrick's Day, in the time of greening rain and wild plum perfume, females arrive if they have not been there already, and spend a few quiet days before their first eggs are laid."

If you want to take a deep dive into the world of screech owls, start with Gehlbach.

Eastern and Western Screech Owls by the Numbers

Height 6 to 10 inches

Weight 5 to 9 ounces, with females larger than males

Wingspan 18 to 24 inches

Life expectancy up to 19 years or more in captivity and up to 14 years in the wild—if they're lucky, as juvenile mortality has been estimated to be as high as 70 percent

Reproduction screech owls achieve full breeding potential by the age of 2 years and fledge two to three young each year

Clutch size three to eight eggs, with four or five more typical

Incubation 26 to 28 days; females may leave the first (and possibly the second) egg unattended for 1 to 3 days before beginning to incubate them

it Kennicott.

The Latin name for the Western Screech Owl, *Megascops kennicottii*, honors Robert Kennicott (1835–1866), who led a three-year expedition to northern Canada and Alaska, where he collected specimens of birds, fish, and mammals for the Smithsonian.

Screech Owl Names

In keeping with standard ornithological practice, we have capitalized the common names of specific bird species but not the group, hence "Eastern Screech Owl" and "screech owl."

The scientific name for the Eastern Screech Owl is *Megascops asio*, and it is a member of the family Strigidae. The bird is sometimes called EASO, bird banders' initials for Eastern Screech Owl. The eastern species is also called the common screech, screech, screecher, shivering owl, little horned owl, dusk owl, little dukelet, and demon owl. We understand most of these nicknames: the bird's whinny call sounds like a shiver, it looks a bit like a tiny Great Horned Owl, and it's seen most often at dusk. (If you know why it's called a little dukelet or a demon owl, let us know.)

The scientific name for the Western Screech Owl is *Megascops kennicottii*. It, too, is a member of the Strigidae. The bird is sometimes called WESO—those darned bird banders' initials again. The western species is commonly known as the mouse owl, cat owl, Puget Sound screech owl, Washington screech owl, and coastal screech owl.

Screech owls, Long-eared Owls, and Great Horned Owls are occasionally called cat owls because of their feline silhouettes.

When "Screech Owls" Aren't Screeches

Writers throughout history have referred to screech owls. But if these writers lived on the other side of the Atlantic, they were no doubt referring to another species of owl.

In ancient Greek mythology, for example, Demeter is said to have changed Ascalaphus, a spirit of the underworld, into a screech owl. The owl in the myth was considered sacred to the underworld gods and a bad omen, but it was more likely a Barn Owl or Pharaoh Eagle-Owl.

An early translation of ancient Rome's Pliny the Elder's commentary says, "The Scritch Owl always betokeneth some heavie newes, and is most execrable and accursed. In summer he is the very monster of the night, neither singing out cleare, but uttering a certain heavie groane of doleful mourning, and therefore if he be seene to flie abroad in any place it prognosticated some fereful misfortune." This creature was accursed perhaps, but not a screech owl.

Shakespeare, meanwhile, referred to "screech owls" a few times, notably in *A Midsummer Night's Dream*, "The screech-owl, screeching loud, puts the wretch that lies in woe in remembrance of a shroud," and in *Henry VI*, in which deepest night is referred to as "the time when screech-owls cry and ban-dogs howl, and spirits walk, and ghosts break up their graves." Again, he was not talking about screech owls.

Leave it to the Hardy Boys to explain. In *The Clue of the Screeching Owl*, a friend of the Hardy brothers drew two owls:

> With amazing skill, Simon sketched in the fierce owl eyes and beak of each. But one of the birds had high-pointed ear tufts; the other seemed to have no ears at all, and had a round, mask-like face similar to that of a monkey.

> "Great sketches," Joe commented. One of his hobbies was ornithology. Now, studying the drawings, he told the others, "The one with the prominent ears is the screech owl. He does the wailing. And monkey face, here, is the barn owl. He does the screeching."

The Hardy Boys knew a screech owl when they heard one.

Screech Owl Distribution

In the Eastern Hemisphere, the equivalent of the screech owl is the scops owl (genus *Otus*), with 58 species. Scops owls have similar habitats, occupy similar niches, and look quite similar to screeches.

At least twenty-five species of screech owls (genus *Megascops*) live in the Western Hemisphere, including the recent discovery of two species of Brazilian screeches. Alas, deforestation threatens both of those species, which have been deemed critically endangered.

North America is home to eight screech owl species: Balsas Screech Owl, Bearded Screech Owl, Eastern Screech Owl, Middle American Screech Owl, Oaxaca Screech Owl, Pacific Screech Owl, Western Screech Owl, and Whiskered Screech Owl. This book deals primarily with Eastern and Western Screech Owls.

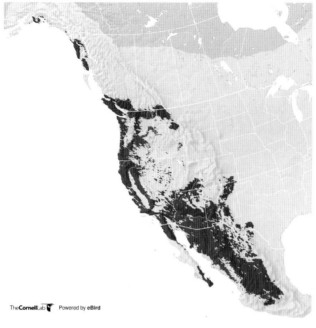

Western Screech-Owl
Megascops kennicottii

Range

The range map depicts the boundary of the species's range, defined as the areas where the species is estimated to occur within at least one week within each season.

▨ Year-round

▨ Modeled area (0 abundance)
▨ No prediction

eBird data from 2006-2020. Estimated for 2020.
Fink, D., T. Auer, A. Johnston, M. Strimas-Mackey, O. Robinson, S. Ligocki, W. Hochachka, L. Jaromczyk, C. Wood, I. Davies, M. Iliff, L. Seitz. 2021. eBird Status and Trends, Data Version: 2020; Released: 2021. Cornell Lab of Ornithology, Ithaca, New York. https://doi.org/10.2173/ebirdst.2020

The Cornell Lab 🦅 Powered by eBird

Eastern Screech-Owl
Megascops asio

Range

The range map depicts the boundary of the species's range, defined as the areas where the species is estimated to occur within at least one week within each season.

▨ Year-round

▨ Modeled area (0 abundance)
▨ No prediction

eBird data from 2006-2020. Estimated for 2020.
Fink, D., T. Auer, A. Johnston, M. Strimas-Mackey, O. Robinson, S. Ligocki, W. Hochachka, L. Jaromczyk, C. Wood, I. Davies, M. Iliff, L. Seitz. 2021. eBird Status and Trends, Data Version: 2020; Released: 2021. Cornell Lab of Ornithology, Ithaca, New York. https://doi.org/10.2173/ebirdst.2020

The Cornell Lab 🦅 Powered by eBird

Clockwise from far left:
Noted author and artist Roger Tory Peterson painted this portrait of *Megascops petersoni*, the South American screech owl species named after him.

Megascops stangiae, a species of screech owl recently discovered in Brazil, was named for Dorothy Stang, a nun who fought for decades to protect the Amazon rainforest.

A Western Screech Owl nests in a cactus.

The Eastern Screech Owl is distributed from Montana and the Great Lakes region east to the Atlantic and from Canada's boreal forests south to the Gulf of Mexico and northeastern Mexico. The Western Screech Owl is distributed west of the Rockies and throughout western Mexico. Their ranges overlap in the Big Bend region of West Texas and a few other locales, where the two species interbreed. Both species tend to mate for life, but the pairs sometimes lead solitary lives after mating season.

Amazing Owls

I used to watch a screech owl ... in a big knothole in a cedar tree.... He stayed in this hole and amazed my cousin and me by appearing one night as a cinnamon-red little bird then amazingly the next night as gray as a weathered old board.

—Angus Cameron, *The Nightwatchers*

Those confusing screech owl color morphs

Eastern Screech Owls are often said to come in two basic shades, reddish brown and brownish gray, but they actually come in a range of colors: a cold gray, a warm gray, brown, a muted orange, a striking orange, and every shade in between. Most Western Screech Owls, meanwhile, have brown or dark gray plumage, with dark streaking on the belly. Ones in the Pacific Northwest tend to be more brownish. In the Southwest, they tend to be more grayish. As with Eastern Screech Owls, the western species comes in a red morph, but these owls are very rare and currently documented only in British Columbia and Alaska. In rare instances, a screech owl's plumage may be white or partly white—the result of a condition called leucism. (To be considered an albino, all the owl's feathers must be white, and the eyes red or pink.)

These two sets of tail feathers show the subtle differences between the Western (top) and Eastern Screech Owl.

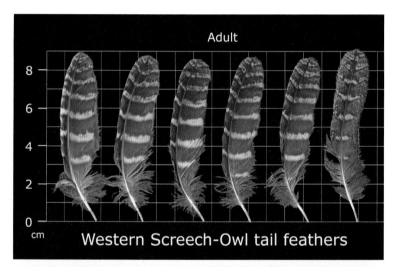

Western Screech-Owl tail feathers

Eastern Screech-Owl tail feathers

When early ornithologists studied screech owls, the red and gray color morphs confused them to no end. Some thought that the owls' feathers changed colors, and others believed that what we now call the Eastern Screech Owl was two species—the red owl and the mottled owl. It's tough to keep track of their theories without a scorecard.

First on the scene was English naturalist Mark Catesby, who made two extended expeditions to the British colonies in North America. Here's how he described "the Little Owl" (*Noctua aurita minor*) in *The Natural History of Carolina, Florida, and the Bahama Islands*, Vol. 1 (1731):

[It's] about the size of or rather less than a Jack-daw; has large pointed ears; the bill small, the iris of the eye of a deep yellow or saffron color; the feathers of its face white, with a mixture

of reddish-brown; the head and upper part of the body a fulvous or reddish-brown color; the wings are of the same color, except they are verged about with white, it has some white spots on the quill feathers and five larger white spots on the upper part of each wing; the breast and belly is dusky white, intermixes with reddish-brown feathers; the tail dark brown, a little longer than the wings; the legs and feet light brown, feather'd, and hairy down to the toes, armed with four semicircular brown talons. The hen is a deeper brown, without any tincture of red.

Swedish naturalist and explorer Carl Linnaeus wrote about the screech owl in 1766, when he described the "Red Owl" in the 12th edition of *Systema Naturae*. In the next volume, in 1788, Johann Friedrich Gmelin described what he called the "Mottled Owl," the gray-morph screech.

In 1812, in the fifth volume of his *American Ornithology*, Alexander Wilson described the "Red Owl" and "Mottled Owl" as distinct species, with illustrations to match.

Sixteen years later, Prince C. L. Bonaparte combined them—sort of. He believed the red-morph bird was the young and the gray morph was the adult. John James Audubon seconded that view in 1832 in his *Birds of America*. In 1837, Dr. S. Cabot Jr. also considered the two birds the same species but made the gray morph the young bird and the red morph the adult. (For proof, he exhibited a mature red morph and young gray that he had shot when they were feeding.)

One would think that the confusion would have ended in 1875, when noted ornithologists Spencer Fullerton Baird, Thomas Brewer, and Robert Ridgway wrote *A History of North American Birds*. They noted that these "two very different plumages are entirely independent of age, sex, or season and that they are purely individual there can be no doubt, since in one nest there may often be found both red and gray young ones, while their parents may be either both red or both gray, the male red and the female gray, or vice versa."

Nonetheless, why some Eastern Screech Owls are reddish and others brownish or grayish still confounded experts. In *Birds of America* (1917), George Gladden wrote that "a captive gray screech owl was fed liver exclusively and its plumage changed to red; but as this was only a single experiment, it cannot be considered conclusive proof that the color change was due to the diet."

Although *phase* was used for the red and gray plumages, Eastern Screech Owls do not change from one shade to the other. A reddish-brown screech owl is

Above left: Eighteenth-century English naturalist Mark Catesby introduced the screech owl to European audiences.

Left and above: John James Audubon and Alexander Wilson initially believed that red-morph and gray-morph screech owls were two distinct species.

Eastern Screech Owl feathers come in shades of gray and red and many shades in between, including this brown morph.

always reddish brown, a grayish-brown screech owl always stays grayish brown, and so forth.

Traditionally, the color variations were attributed in part to geography. Based on studies of similarly plumaged ruffed grouse, the authors of the textbook *Ornithology* postulated that "as the type of forest changes from hardwoods in the south to conifers in the north, and as the depth of winter snow increases, so does the advantage of the gray-phase grouse. Quite likely, this advantage relates to protective coloration and exposure to nocturnal predators."

In fact, a 1963 study found that local Eastern Screech populations vary from mostly red-morph owls in Tennessee to mostly gray-morph owls in Maine, with some intermediate brownish owls thrown in. Both male and female Eastern Screech Owls can be either shade, but it's not quite an either/or proposition. In one Iowa study, females constituted nearly 60 percent of the rufous screeches.

But wait, there's more! According to author and owl expert Scott Weidensaul, recent research suggests the color morphs "appear to confer significant advantages and disadvantages in terms of climate tolerance and even auditory sensitivity." Gray morphs may handle colder temperatures better, and red morphs seem to have better hearing.

Screech Owl Populations

Partners in Flight estimates that 500,000 Eastern Screech Owls and 140,000 Western Screech Owls live in the United States and Canada. Screech owls are far from the most populous owls in North America. That distinction belongs to the Great Horned Owl, with an estimated 6 million birds across the continent. But screeches are the owls most likely to take up residence near you.

Screech owls don't migrate. They live in territories that are about 160 acres, roughly a quarter of the size of Manhattan's Central Park. Owlets start off in their parents' territory, but by summer's end the young get kicked out and must claim their own. By the following spring, they're fully grown, staking out digs in hopes of beginning a new family. That's where you come in.

Heaven knows screech owls could use our help. Many of their habitats have given way to development, with the trees containing their traditional nesting cavities lost in the process. Other threats, including rodenticides and road mortality, persist.

The North American Breeding Bird Survey estimates Eastern Screech Owl numbers have decreased annually by close to 1 percent, for a cumulative decline of about 37 percent between 1966 and 2019. Canada considers the Western Screech Owl threatened, and British Columbia considers it endangered. Several years back, the National Audubon Society selected the Eastern Screech Owl as a species that suburbanites should help, and a key recommendation was to put up an owl box.

A great example of the impact of owl boxes is Grand Junction, Colorado, not known as a screech owl hot spot until 40 years ago. A birder named Rich Levad fell in love with the nocturnal raptors and realized that as the region was being developed, it was losing many of the older cottonwoods where the owls nested.

Levad and the local chapter of the Audubon Society installed owl boxes in favorable habitats and monitored any screeches that took up residence. Fast forward to 2014, when birders in Grand Junction tallied ninety-nine—count 'em, ninety-nine—Western Screech Owls in the proscribed 15-mile radius during the annual Christmas Bird Count. That set a record for the species for the count throughout the United States and Canada, dating back to 1900. The record is especially remarkable because owls are undercounted because they are nocturnal and secretive, and most bird censuses occur during daylight hours.

Equally noteworthy is the largest Christmas Bird Count tally for the Eastern Screech Owl, 237 in a 15-mile radius in Pennsylvania's Upper Bucks County in 2010. That circle still has plenty of heavily wooded areas, but no nest-box program. As one

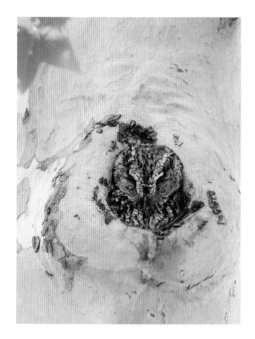

count participant noted, "even smaller, isolated wood lots within semi-developed areas and rural farmed areas can support screech owl family groups."

In short, owl boxes fill a void for screeches in non-rural areas. If your property fits the bill, you may get an owl sooner than you think. In Essex, Massachusetts, an Eastern Screech Owl began roosting in a box within 24 hours after it was installed. Attracting an owl is no sure thing, though. Sometimes getting one to nest in on your property can take years. The good news is that once screeches have found a good spot, they tend to stick around. After they're gone, another screech will likely move in sooner or later, depending on housing supply and existing screech owl populations.

Screech Owl Diet

Without sounding flippant, we'd say that screech owls will catch and devour anything their size or smaller that moves. They hunt from a perch, waiting for prey to pass underneath before pouncing. One YouTube video featured a fearless Michigan screech nailing a Sharp-shinned Hawk—a reminder that although screeches are small, ounce for ounce they rank among the top predators.

SCREECH OWL (Otus asio).

Length, about 8 inches. Our smallest owl with ear tufts. There are two distinct phases of plumage, one grayish and the other bright rufous.

Range: Resident through-out the United States, south-ern Canada, and northern Mexico.

Habits and economic status: The little screech owl inhabits orchards, groves, and thickets, and hunts for its prey in such places as well as along hedge-rows and in the open. During warm spells in winter it forages quite extensively and stores up in some hollow tree considerable quantities of food for use during inclement weather. Such larders frequently contain enough mice or other prey to bridge over a period of a week or more. With the exception of the burrowing owl it is probably the most insectivorous of the nocturnal birds of prey. It feeds also upon small mammals, birds, reptiles, batrachians, fish, spiders, crawfish, scorpions, and earthworms. Grasshoppers, crickets, ground-dwelling beetles, and caterpillars are its favorites among insects, as are field mice among mammals and sparrows among birds. Out of 324 stomachs examined, 169 were found to contain insects; 142, small mammals; 56, birds; and 15, crawfish. The screech owl should be encouraged to stay near barns and outhouses, as it will keep in check house mice and wood mice, which frequent such places. (See Biol. Survey Bul. 3, pp. 163–173.)

With the help of nestcams, we have seen screeches bring in moths, dragonflies, fish, lizards, mice, voles, flying squirrels, bats, and birds as large as flickers. A friend with a nestcam has watched her screeches add frogs, crayfish, and earthworms to the nestlings' menu. The Western Screech Owl's diet can also include such regional specialties as scorpions and pocket gophers.

How does a screech catch a fish? Consider this description from *Birds of America* (1917), edited by Gilbert Pearson: "The screech owl is fond of fish and catches many, especially in winter, when it watches near the breathing holes on the ice, and seizes the luckless fish which comes to the surface."

Perhaps the most unexpected prey was that of a Western Screech Owl that was kept in the monkey house in Vancouver's Stanley Park Zoo in the 1940s. The owl was given free rein of the building, which also contained caged canaries. According to zookeeper F. L. Beebe, "he betrayed the trust and took to killing the canaries whenever he could get close enough to the wire screen to strike. Of course, he could not get them through the screen to eat them, but he never missed an opportunity to kill one." Maybe he was just angry for being held captive.

Owls typically hunt by ear and sight. The jury is still out on their sense of smell. Screeches and other owls will occasionally eat carrion, but dead items are less likely to be *d'entrée du jour* since motionless prey is harder to find. Researchers are using motion-activated trail cameras to better understand how prevalent this behavior is, especially as it may correlate to owls consuming roadkill and getting struck by motor vehicles.

Above left: An Eastern Screech Owl preys on a mouse in this illustration by Ernest E. Thompson (from *Birds of Manitoba*, 1891).

Above: This 1914 screech owl writeup in *Common Birds of Town and Country* talked of the benefits of having these mousers near barns and outhouses.

As for hydration, screech owls drink from ponds and streams. A secondary source of water is from the prey they eat. Nestlings get their water from the prey.

Screech owls sometimes cache their prey, especially during nesting season. In instances where we've had a camera in the nest box and another in an owl box nearby, we've noticed that the owls use the second box as a pantry to store food for crucial times of egg-laying, hatching, and fledging.

Although screeches may sometimes feed small snakes to their nestlings, screech owls in Texas take it to a new level. They import live blind snakes as live-in housekeepers. According to the late Fred Gehlbach, the Roger Tory Peterson of Eastern Screech Owls, these snakes live in the nest debris and eat insect larvae from uneaten prey and other decaying matter. With fewer larvae to parasitize them, he noted, "nestlings with live-in blind snakes grow faster and experience lower mortality."

Screech Owl Habitat

Screeches live in a variety of locales—orchards, wooded suburban areas, urban parks, and wooded areas near streams, wetlands, and meadows. Eastern Screech Owls have traditionally roosted in natural cavities in large trees. The same goes for Western Screech Owls, but they can also nest in holes in large cacti in the Southwest, in keeping with their geography. In recent decades, both species have increasingly adapted to suburban backyards.

Screeches roost in places that offer maximum concealment during the day, preferring conifers, trees with dense foliage, and tree cavities. Screech owls need to blend in with their surroundings for two main reasons—predators and people. Predators include larger owls, mink, weasels, raccoons, skunks, snakes, crows, and jays. Humans are threats because they can't seem to control themselves when they see an owl, especially when they're toting a digital camera or cellphone. No rest for the weary.

If you suspect a screech is hanging out by your yard, scan the taller trees for the owl's silhouette. Owls that lurk in conifers are harder to find. Look for whitewash on branches or the ground under the tree. You might even find a pellet nearby. Just don't disturb the owl—especially if you want it to hang around.

Owls as Fortune Tellers

According to Peter Tate, author of *Flights of Fancy: Birds in Myth, Legend and Superstition*, a belief in India suggested that owls could predict one's fate: "If an owl screeched once, it foretold death; twice, the success of some project; three times, a marriage; four, trouble; five, a journey; six, the arrival of visitors; seven, anxiety; eight, sudden death; and nine, a favourable event."

Screech Owl Calls

Eastern Screech Owls have two mainstay songs: the trill and the whinny. Since screeches are primarily nocturnal, these songs often indicate that you have screech owls in your vicinity. They tend to be most vocal in the late winter and early spring, when pair bonding and courtship are in full swing. During nesting, parents communicate often, mostly in a whisper to avoid attracting attention. They're vocal again in late summer and early fall, a time known as dispersal season, when the parents are reclaiming their territory and owlets are moving out in search of their own.

A trained ear can often tell the difference in pitch between a male and a female. The male's voice is lower, just like in humans. The male's most common song is a mellow trill that lasts 2 to 3 seconds, with about thirty-five notes repeated at various intervals. This rapid staccato has been described as a rapidly repeated *hoo-hoo-hoo*. This trill is also called the bounce song, the A song, the tremolo, the duetting song, or the mating call (when it occurs during breeding season).

The other typical song is the whinny, a series of quivering notes that give little children the willies. This mournful call is made just after dark and typically lasts less than 3 seconds. It is also called the B song. Screeches use the whinny during

The Screech Owl's Call

When other birds are still, the screech owls take up the strain, like mourning women their ancient *u-lu-lu*. Their dismal scream is truly Ben Jonsonian. Wise midnight hags! It is no honest and blunt *tu-whit tu-who* of the poets, but, without jesting, a most solemn graveyard ditty, the mutual consolations of suicide lovers remembering the pangs and the delights of supernal love in the infernal groves. Yet I love to hear their wailing, their doleful responses, trilled along the woodside; reminding me sometimes of music and singing birds; as if it were the dark and tearful side of music, the regrets and sighs that would fain be sung.

They are the spirits, the low spirits and melancholy forebodings, of fallen souls that once in human shape night-walked the earth and did the deeds of darkness, now expiating their sins with their wailing hymns or threnodies in the scenery of their transgressions.

—Henry David Thoreau, *Walden*

Screech owls communicate with a variety of peculiar calls.

courtship and to defend territory. The female may use this song while sitting on eggs, receiving/waiting for prey delivery from her mate, or calling for her mate when he's missing in action.

The two songs sometimes occur in tandem, with one followed by the other. Mated pairs may sing to one other both day and night, depending on the season. Gary Ritchison of Eastern Kentucky University found that "the spontaneous use of bounce songs increased during February and March, declined in April and May, increased again in June, and continued through November. The spontaneous use of whinny songs increased dramatically during August and September and continued through November."

Nestlings *peep* for food during the first 3 weeks, then chatter or hum, according to the Owl Pages website. The female may make odd sounds in response. Fledglings

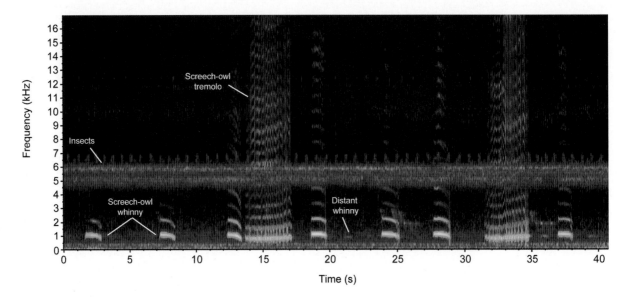

Spectrogram of representative songs of
the Eastern Screech Owl.

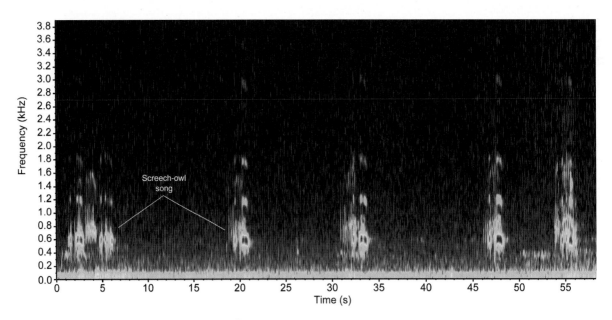

Spectrogram of representative songs of
the Western Screech Owl.

demand food with a harsh *keeeerr-r-r-r*. If you have a nestcam with a microphone, you know how loud and insistent these calls can be.

Eastern Screech Owls have many other calls. Here are a few, courtesy of the Cornell Lab of Ornithology:

Single call notes are used by the male and female during courtship and nesting season. These notes sometimes morph into the trill song.

Loud clacks occur when screeches clap their bills like a movie-set clapboard to signal their annoyance. The owl often clacks when a human is too close to its young. Like the Red-winged Blackbird's sudden fly-by clack, it will get your attention. It's one of the rare times a screech will make a sound in flight.

Chittering is little clicks and other sounds a screech owl makes when agitated or seemingly annoyed.

A three- or four-note rattle or chuckle signifies annoyance, as when a screech is being mobbed by other birds.

Western Screech Owls have two primary songs. The best-known is the bouncing-ball call, which sounds like a ping-pong ball that's dropped on the floor. The second is whistled toots, which begin slowly and then accelerate. The toots can range from five to a dozen, often heard just after dusk.

As Mike Unwin describes the call in *The Enigma of the Owl*: "This repetitive refrain can be heard after dark up the length of western North America. It carries for more than a mile on still nights, providing a nocturnal soundtrack to landscapes as diverse as the humid northwest forests of towering Douglas firs and the stands of saguaro cactus on the edge of the Arizona desert. Rival males contest their territorial claims, each hoping for a female's answer." Once a female arrives, the call becomes a duet.

Western Screech pairs use a short double trill call to keep contact. Females whinny when answering the male's bouncing-ball call and calling for food or copulation. They may make a barking sound or clack their bills when a potential predator approaches.

Spectrograms translate vocalizations into visualizations. If you use the Merlin sound app on your cellphone, you've probably seen them. Since spectrograms for each species are distinct, the app can help identify which bird you're hearing. The app is a handy way to identify the species and visualize the calls when you're listening for owls at night.

The Kay Electric Co. developed the first popular machine for spectrographic analysis in the early 1950s. They called the device the Sona-Graph, and the print-outs it created were known as sonograms. In 1966, *Birds of North America* from Golden Press became the first field guide to include sonograms.

To listen to recordings of Eastern and Western Screech Owl songs, calls, and other vocalizations, visit xeno-canto.org. You can also learn about sounds using Cornell's free Raven Lite app and become more adept at recording birdsong. The free software program lets users record, save, and visualize sounds as spectrograms and waveforms.

Scolding songbirds located this near-hidden screech.

Screech Owls and the Art of Pishing

Screech owls aren't the only ones who like bounce calls and whinnies. Some birders imitate screech owls when they want to attract other birds. The technique is called pishing (no lie). According to ace birder Pete Dunne, who wrote the definitive book on the subject, the purpose of pishing is to tell all birds within earshot that danger lurks.

"My stand-in villain of choice is the Eastern Screech Owl—a very common, mostly nocturnal predator whose daytime presence drives small birds into a vengeful frenzy and whose call, once learned, is both easy to imitate and easy on your vocal apparatus," Dunne writes.

Not only will Western Screech Owls respond to the call, he adds, but Dunne has also used this pish to good effect in Africa and South America: "Maybe it's atavistic. Maybe the calls of Eastern Screech Owl approximate the sounds that small bird-eating dinosaurs used when ordering off the menu."

The Magic of Owls

Anatomy and Behavior

Do you remember the first time you saw an owl in the wild?
Like a first kiss, it's a magical moment not soon forgotten.

Owls have mesmerized humans for eons, inspiring an early artist to limn the likeness of a Long-eared Owl in a cave in the south of France an estimated 30,000 years ago. That artwork still exists. And it's not just any likeness: the owl's head faces front, but the body faces back, making it the earliest known representation of owls' ability to turn their heads 180 degrees in either direction. Many cultures associate that ability with supernatural powers. The owl drawing can be seen as one leaves the deepest sections of the cave, which has raised the question of whether it is using its legendary night vision to investigate the darkness.

Barn Owls are among the world's most captivating.

Ancient Predators

Way before humans, there were screech owls. Researchers in Idaho found 3.5-million-year-old fossils linked to modern-day screeches and other owls. Since fossils of *Homo sapiens* date from a mere 300,000 years ago, that makes us biological toddlers by comparison.

Foremost, screech owls and other owls are raptors. That means they're carnivores, not vegetarians. They have sharp talons and can-opener-like beaks—the better to grab, tear, and eat their prey. They're masters of camouflage, able to blend in with their surroundings and remain invisible when hunting or roosting. And they come in more sizes than any other type of bird—ranging from Blakiston's Fish Owl, which stands 2½ feet tall, to the 6-inch Elf Owl.

All owls have four toes on each foot, three forward and one back. The outer toe on each foot can swivel backward to improve the grip, a trait known as zygodactyl feet. Sharp talons improve their grip as well.

Most owls work the night shift. Roughly 3 percent of all birds hunt at night, and half of those are owls. The other nocturnal birds belong to a family that includes nighthawks, which you may see during migration when they zigzag across the sky at dusk while devouring insects, and Latin America's potoo, a bird that disguises itself as a stick.

The Owl Saves the Day

Because of their physical makeup and nocturnal nature, owls fly under the human radar. That makes having one live next door even more extraordinary. How many of your other neighbors have superpowers? Owls have so many superpowers they even inspired a comic book character named, you guessed it, The Owl.

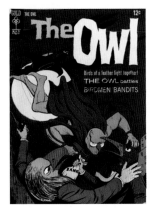

Opposite, clockwise from top left: Owls have four toes on each foot.

This owl drawing in a French cave is estimated to be 30,000 years old.

Potoos blend in perfectly with the trees they perch on.

Common Nighthawks are nocturnal predators with large eyes and camouflaged plumage.

Feathers

Feathers are what make birds unique. No other creature on this planet has them. And no other birds have feathers like owls, not even their fellow birds of prey.

In *What It's Like to Be a Bird*, author David Allen Sibley has a wonderful description of these aerodynamic marvels: "Owls' wing feathers have several adaptations for silent flight: delicate shaggy fringes on the leading and trailing edges, a downy or velvety upper surface, and a generally soft and flexible texture. Softer feathers with more flexible and porous edges allow air to flow more smoothly around the wing, reducing turbulence and therefore noise. These same adaptations also reduce the noise of feathers brushing against each other as the wing moves."

Sibley points out that an owl's body feathers are also soft and fuzzy so that they are silent when they slide across each other. He likens this to "wearing a soft sweater instead of a nylon rain jacket." This has two advantages: "It makes it harder for prey to detect the owl's presence, and it allows the owl to hear any surrounding noises more clearly."

Adult screech owls molt (replace their feathers) annually—in the summer after the fledglings grow more independent. Their feathers get worn, and they replace them through molting. The process takes months, with a few primary flight feathers shed at a time so the owl can continue to fly and hunt. The new feathers, known as pin feathers, emerge from the skin tightly encased in a thin shaft of tissue. The shaft splits, and the new feather opens and grows to its full size. Molting appears to be a key reason screeches use cavities or owl boxes in late summer.

Fledglings of all color morphs have downy gray feathers with darker gray horizontal barring. Only a very subtle warm tint (especially around the facial disks) shows through in owlets that will be red-morph once they lose their downy feathers. As the young owls grow older, they molt their downy feathers, giving way to juvenile feathers that resemble adult plumage but are often paler.

A molting adult owl peering out of a cavity in late summer can sometimes be mistaken for an owlet. You can tell if the owl is an adult in molt or a late-season fledgling by its chest feathers. Owlets are fuzzy gray with only horizontal barring. Vertical marks indicate a molting adult with a fuzzy or even bald head.

In a 1967 study, Joe T. Marshall Jr. posited another reason why screeches molt toward summer's end. That's when their feathers' "evanescent colors and patterns of fresh autumn plumage. . . . are relevant to concealment against the bark and foliage. Worn and faded specimens teach us nothing about nature except that the birds can afford such degradation of pattern in the nesting season."

All screech owl feathers are made of a protein called keratin—the same substance as our fingernails—and come in five basic varieties. Each serves a purpose. As Deane Lewis describes them on the Owl Pages website:

Contour feathers cover the body, wings, and tail.

Soft down feathers provide a layer of insulation next to the bird's body.

Semiplumes fill in between contour and down feathers.

Bristles are small, stiff feathers that occur around the base of the bill and facial disks.

Filoplumes are hairlike feathers covered by other feathers, and they sense the location of other feathers so they can be adjusted properly. The feet and bill have filoplumes that help the owl react to prey and other things they touch.

Although screeches have tufts of feathers resembling ears atop their heads, these feathers have nothing to do with hearing. Known as plumicorns, they are used for displaying and visual communication between owls. The tufts also serve as camouflage by breaking up the owls' outline, making it more difficult for predators to spot them. These feathers also make the owls appear larger when confronted by a predator.

If you watch a screech via a nestcam, you'll notice that it spends a lot of its waking time in the box preening, using its beak and talons to groom its feathers and remove parasites, dust, and dirt.

Above left: The feathers of this gray-morph screech owl allow it to be camouflaged within its nest cavity.

Above: The "horns" on a Great Horned Owl are tufts of feathers used for camouflage to break up the rounded silhouette of the owl's head and for visually communicating with other owls.

Feathers and the Feds

Feathers are generally safe to touch or hold, but it's best to leave them be.

"Most people are unaware that possessing the feathers of most birds is a violation of the Migratory Bird Treaty Act," says Pepper Trail, a former forensic ornithologist with the US Fish and Wildlife Service, former coordinator of The Feather Atlas of North American Birds database, and a leading authority on feathers. "The Fish and Wildlife Service Office of Law Enforcement is not out looking for cases of simple feather possession, but we do investigate cases involving the sale of feathers, which are far more frequent than you may expect."

As far as health concerns associated with having feathers, "There generally aren't any," Trail says. "But that's a moot point since possession is prohibited. So, continue to enjoy, appreciate, and photograph feathers—but don't keep them."

According to Trail, there are a few exceptions: "Turkeys and other non-migratory gamebirds are not covered by the Migratory Bird Treaty Act. And feathers of non-native birds, like ring-necked pheasants, peafowl, and cagebirds like domestic parrots are also legal to possess."

Trail suggests applying for a special permit from the US Fish and Wildlife Service if you're interested in collecting feathers for educational or research purposes. Some states also have permit procedures.

Wonder why it's illegal to own feathers of owls and other birds? Enterprising hunters once killed birds by the thousands for milliners to decorate women's hats—to the point where some egret species almost went extinct. Screech owls were not immune. In one instance in 1911, a millinery store in Detroit was fined $25 for selling screech feathers. The milliner's defense—that the owl came from Japan—did not sway the jury.

In Canada, migratory bird regulations prohibit purchasing, selling, or bartering most feathers. Some families of migratory birds, including hawks and owls, aren't protected under these rules but may be protected under provincial legislation.

Ears and Eyes

Owls use precision hearing to detect the faintest sounds a prey animal makes as it moves through the ground cover. Several physical features augment that hearing. An owl's head can swivel 180 degrees in either direction (although not a complete 360 degrees, as once believed). Because of the tiny time difference in which a sound is heard in one ear and then the other, the owl turns its head until both ears hear the sound simultaneously—indicating the prey is directly in front of it.

To pinpoint the location further, many owls (including screech owls) have asymmetrical ears—one ear is higher than the other—that enable them to determine the vertical location of the sound. Owls have such excellent hearing that they can locate prey under snow or heavy vegetation.

Owls have incredible night vision. They have circular eyes, but they're not like human eyes. Owl eyes are shaped more like tubes, held rigidly in place, and owls must move their entire heads to change their field of view. That's why an owl's head can pivot 135 degrees in either direction and 90 degrees up and down. To help with that flexibility, owls have fourteen neck vertebrae, twice as many as humans.

In addition, owls have huge eyes that account for up to 3 percent of their entire body weight. (By comparison, eyes account for a tiny fraction of a human's weight.) With those large eyes, owls can see great distances and focus on their prey. The only drawback is they can't focus on close objects. Instead, they rely on whisker-like bristles around their beaks to detect those objects.

Unlike most birds, which have eyes on the sides of their heads, owls have eyes that face the front and that—along with flat facial disks for locating sound—sometimes make them appear human-like. Those forward-facing eyes give owls increased depth perception and a greater range of vision.

Owls also can see well after dark, thanks to the unusual composition of their retinas. Retinal cones are responsible for seeing in color and seeing in brighter light. Retinal rods, on the other hand, work best in dim light. By one estimate, rods outnumber cones 30 to 1 in many owl species, enabling them to see well in the dark but not see colors very well. Owls also have a layer of tissue behind the retina that reflects visible light, increasing the amount of light the owl sees and improving its vision at night.

Like many animals, owls have three eyelids. The upper eyelid closes downward when the owl blinks. The lower eyelid closes upward when the owl sleeps. The third eyelid, called a nictitating membrane, is translucent and moves quickly and diagonally from the inner corner of the eye to the outer corner. This membrane helps protect an owl's eyes when catching prey or feeding young.

Opposite, from top: In addition to their feathers, the specialized ears and eyes of owls enable them to hunt at night.

The eye sockets dominate a screech owl skull.

Chapman on Screech Owls

The screech owl frequently makes its home near our dwellings and sometimes selects a convenient nook in them in which to lay its eggs. But its favorite retreat is an old apple orchard, where the hollow limbs offer it a secure refuge from the mobs of small birds which are ever ready to attack it. A search in the trees of an orchard of this kind rarely fails to result in the discovery of one or more of these feathered inhabitants who may have resided there for years. They attempt to escape capture by a show of resistance and a castanet-like cracking of the bill, but when brought from their hiding-place sit quietly, dazzled for a moment by the sudden light. They then elongate themselves and almost close their eyes, thus rendering themselves as inconspicuous as possible. How differently they appear when the western sky fades and their day begins! Is any bird more thoroughly awake than a hungry screech owl? With ear-tufts erected, and his great, round eyes opened to the utmost, he is the picture of alertness.

When night comes, one may hear the screech owl's tremulous, wailing whistle. It is a weird, melancholy call, welcomed only by those who love Nature's voice whatever be the medium through which she speaks.

—Frank M. Chapman, *Handbook of Birds of Eastern North America*

Sense of Smell

For the longest time, the conventional wisdom was that owls have a poor sense of smell. Sure, turkey vultures could smell carrion from great distances, but what about the Great Horned Owl? If it could smell, why would it prey on foul-smelling striped skunks when all other animals avoided them?

Danielle Whittaker, the author of *The Secret Perfume of Birds*, concluded otherwise about owls' sense of smell—beginning with the false assertion that Great Horned Owls were the skunks' only predators. According to the evolutionary biologist, it's not that birds lack a sense of smell, it's that their sense of smell differs from ours. During her research with Dark-eyed Juncos, Whittaker learned that birds use their sense of smell for picking mates, finding their nests, and other purposes.

Owls and other birds employ scent as well. For example, owls—especially larger ones—may eat carrion more often than previously thought, as evidenced by images from motion-activated trail cameras. As for screeches contending with the stench inside a late-season nest box? Perhaps they just smell things differently.

Intelligence

Throughout history, owls have symbolized wisdom. The ancient Greeks, for example, believed that Athena, the goddess of wisdom, shifted back and forth between human and owl form. Many New York City school buildings built in the late 1800s to early 1900s featured carved-stone owls above the main entrance. Wise Potato Chips have an owl logo. And while collective nouns for other birds range from a gulp of cormorants to a puddling of mallards, owls qualify as a parliament.

But just how smart are they? After all, we know ravens can do all sorts of cognitive tricks—a recent study found that young ravens are as smart as many adult primates in terms of problem-solving, spatial skills, and learning abilities. Brown Thrashers are avian Spotifys, able to sing up to 2000 songs. Tiny titmice and chickadees remember where they have stored thousands of individual seeds. Blackpoll Warblers can fly 1800 miles over the Atlantic and still find their breeding grounds. All these species belie the expression "bird brains."

As Jennifer Ackerman writes in *The Genius of Birds*, "There are birds that can count and do simple math, make their own tools, move to the beat of music, comprehend basic principles of physics, remember the past, and plan for the future."

What, then, of owls? The answer depends on how you gauge intelligence. Ask a search engine if owls are wise, and you'll get lots of speculation. The problem lies not with the answers but with the question. As in most instances, when humans gauge the traits of other animals, the definition of intelligence is in the eye of the beholder.

"Parrots and many other species are smarter than owls," says Wayne Lynch, author of *Owls of the United States and Canada*. "The wisdom of owls is just an old myth."

On the other hand, Tony Angell, author of *The House of Owls*, says, "We can never begin to fathom the cognitive complexities of other species that are fashioned for lives far different than ours, but we can at least appreciate or imagine the infinite possibilities that exist."

Angell points out that when a pair of Western Screech Owls nested near his home, they did not seem disturbed by his presence near the owl box. Angell says, "That demonstrated the owl's capacity to identify individuals. I had established myself as a non-threatening human."

Most owl behaviors, from selecting a nest site to caching food to protecting fledglings, involve brainpower, be it hard-wired or otherwise. By human yardsticks for intelligence, screeches are nowhere near as smart as we are. But at the rate we're going, screech owls might outlive us all. Maybe *Homo sapiens* aren't as *sapien* as we'd like to think.

Pellets and Whitewash

Although all raptors and even many non-raptors such as kingfishers also cough up the indigestible parts of their meals, owls have become synonymous with these furry pellets. Actually, owls expel their food waste in two ways: pellets and whitewash.

Since owls lack teeth and cannot chew their food, they often consume smaller prey whole, including feathers, fur, teeth, and even bones. The owl's gizzard compresses these indigestible parts into pellets that the owl then regurgitates after several hours. Depending on the type of owl and its digestion, pellets can range in length from a ½ inch to 4 inches. Newly ejected pellets are moist and black, slowly turning gray as they age. They're typically found near the owl's nest or favorite roost.

The rest of the owl's waste, a white semi-liquid known as urea but more typically referred to as whitewash, is ejected from the vent, a reproductive and excretory opening located under the owl's tail. The nitrogen-rich urea is similar to mammal urine, except white and thick.

The pellets that educators and amateur naturalists typically dissect are those of Barn Owls. They are easily collected, relatively large, and usually contain mice. The business of collecting and processing Barn Owl pellets for dissecting kits even made it onto a TV episode of "Dirtiest Jobs."

If you want to go that extra educational step of collecting and dissecting pellets, there are lots of quality free online curricula (such as the Cornell Lab of Ornithology's) to get started identifying the owls' diet. Raw pellets can carry diseases, so precautions apply. To disinfect the pellets, wrap them in aluminum foil and bake in an oven at 325°F for 40 minutes. Fair warning: the aroma won't be delectable, so you might want to use a toaster oven in a well-ventilated area or take it outside.

As for what becomes of the nestlings' waste, the jury is still out. The young of several cavity-nesting birds, including the Eastern Bluebird, excrete their waste within a mucous membrane known as a fecal sac—a whitewash sausage or an avian diaper, if you will. A parent (most likely the female) then consumes it or discards it well away from the nest. We suspect this is the case with screech owls as well. If you know for sure, please contact us.

Opposite, clockwise from left: A Great Horned Owl coughs up a pellet.

Owl pellets come in various sizes and colors: (left to right) Eastern Screech Owl, Barn Owl, and Great Horned Owl pellets.

Bluebirds remove nestlings' waste via fecal sacs. Screech owls likely do this as well.

Protecting Owls

Throughout recorded time, owls have been considered omens, most often of death and bad luck. That's not a big surprise, considering how they live under the cover of darkness, fly silently, and make eerie sounds in the dead of night. That has put a metaphorical target on their backs.

Take this letter to the *New England Farmer* back in 1860: "The owl robs your henhouse in the blackest night. Educated thieves go to jail for the same offense. Which of the two are better members of society?"

Shooting owls was legal in many states—and often encouraged. By the early 1880s, farmers and others saw hawks and owls as a Public Enemy No. 1. Legislators in several states passed incentives for people to kill them, and in some cases the bounties lasted into the 1930s.

This Eastern Screech Owl was found as a nestling and raised in captivity by well-meaning people who were unaware that raising captive wild birds is illegal and that the owl would become imprinted on them and could not return to the wild. He now lives at The Raptor Trust in New Jersey.

SCREECH OWL.

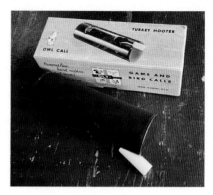

Above: This owl call was mass produced in the 1950s.

Left: *Birds That Hunt and Are Hunted* (1898) noted that people in cities were most familiar with screech owls "on women's hats, worn entire or cut up in sections."

Pennsylvania's *Altoona Times* announced in 1885:

> All the farmers and farmers' boys in this Commonwealth should know that on the 23rd of last June the Legislature passed a law for the benefit of agriculture and the protection of game, by offering a bounty of two dollars for every wild cat killed, one dollar for every red or gray fox slain, and fifty cents for every mink, weasel, hawk, or owl, excepting the screech or barn owl. Persons who kill any of the above, and desire the reward, must go before a Justice of the Peace with the slain bird or

animal and make an affidavit of the time and place of the killing, when an order will be drawn on the county treasurer for the amount.

Even after the bounties went out of fashion, the mind-set persisted. A mass-produced "turkey hooter" and "owl call" from the 1950s came with instructions that advised the call "can be used to call owls and for shooting the owls in the states where the law does not prohibit."

In 1972, amendments to the Migratory Bird Treaty Act of 1918 finally included raptors and made it illegal "to pursue, hunt, take, capture, kill, possess, sell, purchase, barter, import, export, or transport any migratory bird, or any part, nest, or egg of any such bird."

Owl Superstitions

As with owls all over the world, screeches have traditionally been associated with death, wisdom, and sorcery. This holds true across most North American civilizations, from the Aztecs of Mexico to the Tlingit people of Alaska, who told the tale of how a cruel woman was turned into the first screech owl after burning her mother-in-law with hot coals.

"One generalization, which may be true for quite a few different tribal nations across the continent, is that the call of the screech owl might be a message of impending death or disaster," says Abenaki writer and traditional storyteller Joseph Bruchac, best-selling author of *Keepers of the Earth* and other children's books. "Or a message that someone already died."

Is this Western Screech Owl giving you the evil eye?

The Caddo Native Americans of the Southeastern United States believed that witches could become screech owls or become partners with them. After bewitching, the screech owl would be sent by the witch to an invalid's house to spy on how he is doing. Accordingly, whenever people would see a screech owl about the house, they would try to kill it. Whenever a bird was shot, there would be a hole or bruise in the corresponding part of the witch's body—akin to an avian voodoo doll. If the owl was killed and boiled in a kettle, the witch would die.

The ancient Cherokees had several superstitions involving screech owls. Some chiefs were said to foresee the outcome of an upcoming battle by screech owl calls. Cherokee mothers sometimes hushed crying children by telling them that a screech owl was listening in the woods. Cherokee shamans valued Eastern Screeches as consultants and claimed the owls could bring on sickness as punishment.

According to *The Journal of American Folklore*, screech owls inspired many nineteenth-century superstitions, including:

- To make a screech owl leave, tie a knot in the bed sheet.

- Old maids turn into screech owls.

- A screech owl hooting near your window is a sign of death or bad luck.

Screech superstitions persisted well into the twentieth century. As Edward H. Forbush wrote in 1927, "The little screech owl of the South is known in Louisiana as the 'Shivering Owl'; and along Bayou Lafourche when its notes banish sleep, and the resourceful 'Cajun' wishes to ward off the ills that he believes otherwise sure to follow, he must arise from his couch and turn his left shoe upside down. Then the cries are supposed to be stilled."

Forbush reported that even in prim and proper New England, screech superstitions persisted: "One that took up residence for a few days in a church tower was credited with foretelling—if not indeed causing—the death of a citizen of dignity, domiciled next door."

Thankfully, the old superstitions and legends about screeches have become increasingly anachronistic to the point where screeches have had their moment in the sun—and not just perched in the opening of their nest box.

The Aztecs considered horned owls to be omens of death. This image is from the *Florentine Codex Book V*, 1577.

This Tlingit bowl, carved in the mid- to late-1800s, features owls carved into both handles.

An Eastern Screech chills at the owl box opening.

Screech Owls and Egg Collectors

While states were offering bounties on owls and other wildlife, ornithological supply companies were selling bird skins of every feather, along with the eggs of hundreds of species. *The Ornithologists' and Oologists' Manual* (1895) was actually more of a catalog of birds, preserved bird skins, and hollow bird eggs for collectors. Sample prices: a bald eagle skin sold for $8, a screech owl skin sold for $1, and a Kennicott's (western) screech owl skin sold for $2. The skin of the likely-extinct ivory-billed woodpecker sold for $15—making it more valuable to kill and collect than other species.

It is hard to believe nowadays, when collecting eggs is heavily regulated, but for several decades during the Victorian era, bird-egg collecting was the rage. Thousands of amateur oologists collected millions of wild bird eggs all over the world, and nobody thought anything of the practice. Some even saw it as a way to get people, especially youngsters, to experience nature and the great outdoors.

If you wanted to collect screech owl eggs without robbing a nest, you could buy them from egg suppliers. The 1885 edition of *The Oologist's Handbook* sold eggs ranging from 60 cents for a generic "little screech owl" egg to $1.50 for a "Northwestern screech owl" egg. (A US dollar in 1885 was the equivalent of roughly $30 in 2022.)

Screech owl nests were a favorite target for some. If you go back and read the reports from collectors in 1921 editions of *The Oologist*, a monthly publication "for the student of birds, their nests, and eggs," you'll find these antiquated reports:

> "Most field men know that many birds will, if their eggs are taken, at once nest again. Owls will for at least three times. . . . I have taken screech owl sets of five and within two weeks more a set of two well incubated."

> "Thinking that my experience with the little screech owl will interest some readers of *The Nidiologist* [a periodical for those interested in nesting birds], I write a few lines showing the tenacity for locality of this little hooter. . . . We found Mrs. Owl at home and two almost round glossy white eggs beneath her— this was April 15. I paid her a visit on the 19th and collected a set of four eggs, and also collected the old lady, thinking I would have her mounted."

"Thirty years ago today (March 20, 1891) from a cavity in a large black oak . . . I took a set of five screech owl eggs. About two weeks later I took another set from the same cavity. Both sets are in my collection at the present time."

For do-it-yourselfers who collected their own screech owl eggs and other bird eggs, one egg-supply house offered oologists' outfits. The basic kit, which sold for 25 cents postpaid, included a white metal blow-pipe, a small drill bit, a nickel embryo hook, a lead pencil, and a 5-inch ruler, "all in a neat box."

The Need for More Citizen Scientists

For centuries, screeches and other owls were the subjects of all sorts of myths, superstitions, and misinformation—and often killed as a result. Until recently, changing those perceptions was next to impossible because screeches are secretive and nocturnal birds, out of sight of human observers. Much of what has long been considered basic knowledge about screeches was based on assumptions, and those assumptions have often been wrong.

One of the most striking aspects of our research for this book was how often the information we sought was contradictory or nonexistent. After all, the first organized citizen-science effort to compile information on screech owls did not begin until the 1960s, when bird-watchers from across the United States and Canada observed the nesting behavior of their local birds, recorded the information on index cards, and sent them to Cornell University.

Over the ensuing four decades, the Cornell Lab of Ornithology's North American Nest-Record Card Program collected more than 300,000 records on bird

Opposite: Screech owls nest in tree cavities, when available.

Below: A 1986 record of a Western Screech Owl nest from the North American Nest-Record Card Program.

nests from citizen scientists, including ten records of Western Screech Owl nests. The sample size, spread over five states, provided a snapshot. The screeches nested about equally in Wood Duck boxes and tree cavities—plus two nestlings that were discovered in a hole in a telephone pole that had just been cut down.

Although the almost perfect survival rate of the young owls noted on the cards was excellent, it likely did not reflect the bigger picture. Because observers seldom looked inside the nest cavities, they never saw how many eggs went unhatched or how many nestlings died before they could fledge.

Similarly, before the advent of nestcams, naturalists thought that screech owl incubation began with the arrival of the first egg. When average citizens were able to document what happened inside the nest boxes, however, they noticed that the eggs were often left unattended overnight until two or more eggs had arrived, confirming that incubation began much later than had been assumed.

Researchers' understanding of screech owls continues to improve with ever-improving technology available to all, with an assist from those old nest-record cards. That old data, combined with hundreds of more recent sightings filed online by citizen scientists with nestcams, provide researchers with an unprecedented knowledge base about owls and other birds.

"What these personal observations show is that the screeches are adaptive and resilient and that they can thrive alongside humans," says Cornell's Robyn Bailey.

If you have a nestcam, you can provide all sorts of nesting data to Cornell's NestWatch.org. This includes the number of eggs, when they arrive, when they hatch, and whether the nestlings fledged. You can also submit your observations via NestWatch.org apps for Android and IOS.

"It's not citizen science until you submit that data," Bailey says. "It's a great way to close the loop and get that information into the hands of the scientific community."

Gigantic Nest Boxes to the Rescue

More and more owls are nesting in boxes. One surprising example is on Japan's Hokkaido Island, where Blakiston's Fish Owls nearly went extinct in the 1990s. One main reason for their decline was the loss of enormous poplars where these huge, endangered raptors nested. Once researchers realized this, they made nearly 200 nest boxes from 55-gallon plastic barrels and placed them 35 feet up in trees. These days, 80 percent of Hokkaido's fish owls—made famous in *Owls of the Eastern Ice*—rely on nest boxes to breed.

The survival of Blakiston's Fish Owl, the largest owl in the world, was assisted by the use of gigantic artificial nest boxes.

Imagine, gigantic nest boxes to save 10-pound owls. What would the right owl box in your backyard do?

Owl-Friendly Yards

You can help owls and other wildlife by making your yard more nature-friendly. Improving your yard and your home's exterior to make your yard friendly to screech owls is as important as taking steps to keep squirrels and other predators from invading owl boxes. Just as new parents go around making sure that cabinet doors have latches and all stairs have safety gates, you should look for potential owl hazards and eliminate them as best you can. These simple steps not only help owls, but they also help pets, insects, and other birds and wildlife.

We love the way Marlene A. Condon describes what a healthy yard should be like in *Nature-Friendly Garden*: "The mostly unrecognized truth is that our yards and gardens need to function in much the same way a wilderness area does. A gardener who obeys the laws of the natural world owns a garden that is home to numerous organisms (predators and prey) that keep it functioning properly and trouble-free." Vegetable gardens and compost piles attract prey items for screech owls. Although attracting rodents to your yard may seem counterintuitive, screech owls will thank you.

Consider planting native shade trees that can provide hunting perches over open areas. Evergreens with dense branches provide places for owls to roost. While you're at it, you might rethink that chemically treated lawn, if you've got one. Clovers and dandelions are beneficial to wildlife. Even those beetle grubs that can turn a yard

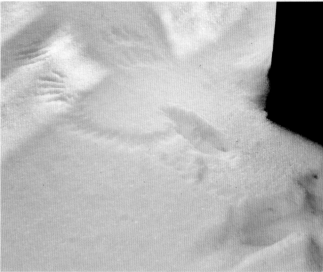

brown provide food for many birds and mammals. (If you need expert advice, ask the nearest native plant society.)

Trim your shrubs and cut down trees (if you must) anytime except in the spring. Chris Soucy, executive director of The Raptor Trust, a leading bird-rehab center in the New York City area, says, "We see a lot of baby screech owls in spring and summer displaced from nests when people do spring cleaning." Soucy recommends that people always do that kind of yard work in the fall—after nesting season. "The same is true for shrubs and bushes," Soucy says. "Can't even count how many baby catbirds we get each year from shrubs people have trimmed the nest right out of."

Similar projects near an occupied nest box in the fall or winter may cause the owls to relocate, usually temporarily, but Soucy advises against going overboard when trimming trees. If a dead tree (a snag) isn't in danger of falling on your house or garage, consider leaving it standing. Screech owls love to perch and roost there and sometimes nest in old flicker or Pileated Woodpecker holes.

Install a birdbath. The National Audubon Society advises that screech owls drink and bathe in shallow water daily. "Adults and fledglings habitually use suburban birdbaths, a behavior [that] may go largely undetected by humans because it generally occurs at night." Another bit of advice they offer is to "keep water clean and free of mosquito larvae."

Bird feeders may increase songbird mobbing of the screech parents and nest box, which may be less than desirable for an owl, but fallen birdseed also attracts nocturnal rodents that screeches appreciate. Although screech owls will prey on

Above left: A Western Screech Owl hangs out in a yard in Colorado.

Above: If you look closely, you can see the impression in the snow left by an Eastern Screech Owl mantling its prey near a compost bin.

songbirds, it's usually between dusk and dawn, picking them off from their roosts. Screech owls seldom hunt at bird feeders during the day but will occasionally if they're very hungry or they have owlets to feed. More likely, they will go after mice attracted to the feeder after dark. The bigger concern with feeders is the squirrels—nest-box squatters—they attract.

Clean off the road in front of your property. This might sound like a head-scratcher, but a leading cause of death of screech owls and other raptors is flying into vehicles or getting hit by them. One reason is the apple cores, half-eaten bananas, and other garbage the people toss from their car windows with the rationale that food is biodegradable. The discarded food attracts various rodents, which in turn attract owls and other raptors to the side of the road, where they can collide with vehicles.

Keep cats indoors. You may think of Mittens as a cuddly house cat, but she is a predator at heart. Not only will she attack small birds for sport—even screech owls and their young—but she'll also eat other items in the screech diet, such as mice and lizards.

You should also keep Mittens inside for her own protection—from eating a poisoned rodent, getting hit by a car, or getting nailed by a Great Horned Owl after dark. According to the U.C. Davis School of Veterinary Medicine, the average house cat's lifespan is 10 to 15 years, while the typical outdoor cat may only live 2 to 5 years due to the dangers it faces.

Check Your Chimney

Almost everyone has heard a story or two about a duck flying down a neighbor's chimney. Turns out that screech owls accidentally fly down chimneys as well. A homeowner in Iowa found a scorched screech owl on her lawn the day after she lit a fire in her fireplace. In another instance in Tennessee, a screech got stuck in a stovepipe flue—twice. In both cases, the homeowner caught the owl in a towel and put it outside. The owl finally realized that trying again would be a bad idea.

If you have a chimney with a metal liner, put a cap on it to keep out rain, birds, and other wildlife. As Georgean and Paul Kyle, project managers for the Chimney Swift Conservation Association, explain, "Chimneys with metal liners should always be capped because they are traps for any animal that enters."

If you have a masonry chimney, proceed with caution—especially if resident or migrating Chimney Swifts are in your area. In those instances, use a chimney cap that still enables these birds to nest in the spring and/or roost in late summer.

According to the Kyles, "An alternative to a permanent cap is to have a professional install a cap during the burning season (when the swifts are not in North America) and then have them return to clean the chimney and remove the cap for the late spring and summer nesting season. Then replace the cap in the fall. This annual maintenance routine is good for the swifts and homeowners alike."

Reduce the Pain from Windowpanes

One of the most sickening sounds that bird-lovers will hear is that unmistakable *thonk* of a bird crashing headlong into one of their windows. Unfortunately, that *thonk* is all too common. Researchers estimate that 160 million to 440 million birds—including screech owls—die in the United States each year from flying into residential windows.

Early one February morning in 2021, Andy Bunge of Preston, Minnesota, was sitting by his picture window, when he heard that sickening thud. He knew immediately what had happened. A bird must have flown into the window, but what bird would be flying this time of day?

He looked down and saw a screech owl lying dazed in the snow. He picked up the dainty owl and went into action. The local owl center connected him to the University of Minnesota Raptor Center in St. Paul. He sent a photo of the owl to the staff there, who immediately saw an injury above the left eye. They asked that Andy help get the owl to the center—a two-hour drive—for a closer look. The staff X-rayed the owl and treated its injuries, and it was released a week later in Andy's neck of the woods.

"You can tell a lot about people by the way they treat nature," notes Andy. "In this day and age, we get so caught up in technology and life in general that we overlook our relationship with animals in the wild. How do we coexist in a way that allows these species to prosper? We have to be more sensitive to the creatures around us."

Daniel Klem, the leading expert on bird-safe windows and author of *Solid Air*, the go-to book on the subject, says that while owls are considered nocturnal, they often hunt at dusk and dawn when a clear or reflective window can fool them—often lethally.

Klem's advice:

- If you have window screens, keep them up year-round. Screens keep birds from striking the unyielding glass surface.

Andy Bunge found this dazed Western Screech Owl below one of his windows.

- On windows where bird strikes have occurred, apply protective window patterning such as Feather Friendly window markers a few inches apart. If you remodel, consider new bird-safe glass.

- Hang Acopian BirdSavers (parachute cords) on your larger bird-prone windows. The company sells BirdSavers kits and offers free instructions for do-it-yourselfers.

Klem says that while keeping your windows dirty might work in theory, they would have to be too dirty to look out of to be effective.

What should you do if a screech owl flies into your window and is too stunned to fly away? The West Virginia Raptor Rehabilitation Center, which treated a window-injured screech in 2020, advises:

> Using a towel, gently place the bird in a box. Put the box in a dark, quiet place away from pets or other predators. Check on the bird every half hour or so to see how it's doing. Do not open the box inside your home—take it to a place where the bird can easily fly away if it's ready to go! Do not give it food or water, and try not to handle it more than necessary.

Birds usually revive quickly after a collision and can be released soon afterward. If the bird shows signs of injury or does not revive in a couple of hours, call your local rehabilitation center.

Whenever dealing with a stunned screech owl, wear thick leather gloves if possible. An unconscious screech owl is no problem to pick up, but that same bird can inflict serious damage with its sharp talons when it becomes alert.

For the name and phone number of a raptor rehab center nearest you, go to "Helping Birds" in the "Birding for Beginners" section of Birdwatchingdaily.com and click "Raptor rehab centers in the US and Canada."

When Bad Things Happen

Unfortunately, The Raptor Trust's Chris Soucy has plenty of expertise on screech owl injuries and calamities. His center treats forty to fifty screech owls a year.

"The most common reason for admission is impact injury," says Soucy. "Screech owls get struck by cars regularly. Roadways in wooded areas also make attractive flyways for birds. They hunt the roadsides for prey, but it's a dangerous place."

Experienced rehabilitators often use avian foster families when possible. In 2021, a raptor rehabilitator we know received an orphaned screech owlet. She connected with someone who had a screech box and nestcam. The nest had three owlets roughly the same size as the orphaned one. The rehabilitator placed it with the other three, and the parents accepted it. All four fledged a little more than a week later.

If screech owls roost or nest nearby, the time may come when you find an injured screech owl or a displaced owlet. Here's Soucy's advice:

What should you do if you find a nestling on the ground or displaced from its nest?

The best course of action is to contact the nearest wildlife rehabilitator. That doesn't necessarily mean picking up the bird, putting it in a box, and transporting it to a wildlife center—yet. It may come to that. But some birds can be placed back in their nests, and allowing their natural parents to raise them is the best course of action if it is possible. If they are old enough, some juveniles can even be left where they are. Every set of circumstances is unique, and it's not quite as simple as, "I found this baby bird, so I'll stick it back in this nest." A licensed wildlife rehab professional can help assess the situation.

What should you do if you find an injured screech owl?

Screech owls and all other native wild birds in North America are protected by federal laws. Injured birds should be taken to a licensed, trained wildlife rehabilitator for care. Sometimes that seems impossible, but there are many resources available. State Fish and Wildlife agencies keep lists of licensed wildlife facilities and can help people locate the nearest one. Local animal shelters and animal control officers may be able to help transport injured animals to rehab centers, and in many cases they can assist with capturing injured animals that have been found.

If you need to pick up the owl, do you need special gloves?

Screech owls are small compared to many other raptors, but they do have sharp talons and beaks. Lightweight leather gloves or garden gloves can protect your hands. A lightweight towel can be tossed over a screech owl so it can be scooped up.

What sort of carrier should you use if you bring it to a raptor center?

If one is available, a soft carrier is safest for transporting wildlife—the kind that might be used to bring a cat onto an airplane, for instance. Screech owls are wild animals, and they don't know that a person may be trying to help. They will try to escape, and they may thrash about in a carrier.

If a soft carrier is not available, a cardboard box with a towel on the bottom will work. Metal bird cages like those used for parakeets and other exotic pet birds are not recommended—they are very rigid and wild birds can injure themselves inside of them. If the bird seems agitated, a sheet can be draped over the top of the carrier to help keep it calm. Warm, dark, and quiet is the best possible temporary environment for an injured or orphaned wild animal.

Beware of -Cide Effects

Avoid using any product in your yard that ends with the suffix -*cide* (from the Latin *cida*, meaning "killer"). These include rodenticides, insecticides, fungicides, and herbicides. These poisons are not nearly as specific as their manufacturers would lead you to believe.

The most dangerous to screech owls and other raptors are anticoagulant rodenticides. These poisons work by blocking the rodent's ability to process Vitamin K—no Vitamin K, no ability for blood to clot. That causes the rodent to bleed to death. This can take several days, making the weakened rodent an inviting prey for owls, other wildlife, and large pets.

Not long ago, a raptor rehabilitator in Gloucester, Massachusetts, took in an Eastern Screech Owl that was in rough shape. The staff drew a blood sample. After 30+ hours the owl's blood still had not clotted, confirming poisoning by rodenticide. Treatment was an injection of Vitamin K every 4 hours, day and night, for a week. As the owl's health improved, the frequency of the injection could be decreased. After about 5 weeks they stopped treatment to see if the owl could keep its Vitamin K levels up. The blood clotted in less than 90 seconds, and the staff released it into the wild a few days later. This one was lucky. Most don't survive.

Other household rodenticides aimed at mice and chipmunks typically contain the first-generation anticoagulants warfarin, chlorophacinone, and diphacinone. These poisons can take up to 2 weeks for rodent mortality to occur.

When a raptor eats a poisoned rat, it can consume a lethal dose of the poison as well. The irony is that the predator could otherwise consume more rodents than poisons would kill over the course of a year—without those "cide" effects. Raptors Are The Solution (RATS) and saferodentcontrol.org are two initiatives to educate public and private property owners and managers on the benefits of raptors versus poisons in rodent control. In a perfect world, pest-control businesses would provide owl boxes to their customers.

In December 2020, a sickly Great Horned Owl died en route to The Raptor Trust in New Jersey. Because the owl showed no external signs of trauma, a necropsy was performed to determine the cause of death: rat poison. Since then, the body count of raptors killed by rat poison in that vicinity has climbed by more than half a dozen.

If you live near a restaurant, warehouse, or office building, also look out for gray or black boxes known as rodent bait stations. They dispense stronger (four- to five-day mortality) anticoagulants such as brodifacoum, bromadiolone, difenacoum, and difethialone, which have been banned for household use and are even more lethal to raptors and other wildlife and pets; these are often referred to as SGARs, second-generation anticoagulant rodenticides. California and British Columbia have placed a moratorium on the use of the deadliest poisons.

With the increasing regulations on anticoagulant-type rodenticides, household rodenticides with either cholecalciferol, bromethalin, or zinc phosphide are increasingly prevalent. Cholecalciferol is activated Vitamin D. This may sound harmless,

Black-box poison dispensers abound in downtowns and office parks.

This Great Horned Owl died on the way to a bird infirmary. The cause of death was rat poison.

Household rodent bait stations contain poisons that can kill owls and other raptors that eat the poisoned rodent.

Household insecticides kill more than ticks and mosquitoes. Butterflies, moths, and honeybees are vulnerable as well.

but in concentrated amounts, it produces the buildup of calcium and phosphorus that causes acute kidney failure. Bromethalin is a nerve agent that causes respiratory arrest, and zinc phosphide produces highly toxic phosphine gas in the stomach when ingested.

Currently, the consequences on raptor populations are not as established with these poisons as they are with the anticoagulant types. For owls, however, no rodenticides are likely safe.

For wildlife and pets, the best-case scenario for the household poisons is the rodent dies inside a wall where it won't be consumed. The dead rodent will produce a nasty odor and plenty of flies in the building, two more reasons to avoid using these poisons.

Other pesticides kill all sorts of insects, including butterflies, moths, honeybees, and dragonflies. Pesticides lethal to large insects that owls eat in the summer may poison the owls as well, and herbicides often kill plants that are beneficial to those insects.

"Screech owls are able to prey on a diversity of animals and these chemicals can have adverse effects on their food supply," says Annick Gionet Rollick, senior wildlife rehabilitator for The Owl Foundation. "Screech owls have been known to prey on moths, June bugs, worms, mice, birds and even amphibians, reptiles and fish." In short, while these poisons target one link in the food chain, they affect the entire chain, including the local groundwater.

Go on an Owl Prowl

Environmental education programs are vital in conservation efforts across the world. If you've never looked or listened for an owl in the woods at night, you should—the first chance you get. One of the best and simplest ways is to participate in an owl prowl. Environmental groups across the continent typically hold these bird-watching events in the winter to introduce the public to the wonders of nature at night. Owls make for magical ambassadors simply by being themselves.

Just ask birder Cliff Hagen of Staten Island, New York, a full-time special-ed teacher and part-time naturalist who has led owl prowls for two decades and introduced more than a thousand people to the wonders of nature at night in the process.

Hagen thinks of owls as neighbors he wishes he could get to know better. "They are in nearly every park on Staten Island, just beyond our yards," he says. "Yet so few people ever get to experience the rewards of a night visit with our nocturnal, woodland neighbors. The whistle of a screech owl is meditative, and the resonance of a Great Horned Owl's call is captivating."

Listening to a recorded owl's vocalizations can't compare to hearing a real-live owl after dark. It is, quite simply, the genuine article—the call of the wild.

Our First Owl Experiences

I can't say this was the first owl I ever saw, but it was one of the most memorable. When I was in fifth or sixth grade, I went on an owl prowl at 3:30 a.m. for the National Audubon Society's annual Christmas Bird Count. Our leader did a few owl imitations, and I was fascinated.

Great Horned Owls at times were calling unprompted in the distance, but it was the leader's screech owl imitation—a combination of whistling and gurgling—that attracted a handful of screech owls close enough that we could see them in the dark.

By the time we reached the last stop, my toes were ice-cold and a-tingle, and 11-year-old me was content to sit this one out in the car. But I couldn't leave well enough alone. I decided to practice my own screech owl imitation. Little did I know cars aren't that soundproof in the quiet of the early morning.

The leader appeared abruptly and inquired if I had been imitating screech owls. I nodded. He gave me a look and a restrained "Please don't."

My stomach churned, but it was a lesson learned: The excitement of seeing or hearing a real owl is nothing to be trifled with.

—Scott Weston

I regret that it took me well into adulthood to see a live owl—and a screech owl at that. My wife and I had just moved into a home next to a nature preserve, and one afternoon the marsh warden called to ask if we'd like to see a screech sunning itself in his yard. We jumped at the chance and soon found ourselves tiptoeing around the corner of his house. He pointed to an ash tree and whispered, "Right there."

I could see nothing until the marsh warden said to look at the eye-level knothole. When I looked more closely, the owl materialized, its gray feathers blending with the bark. To see something so wild and so mysterious so close was transformative.

—Jim Wright

GARDNER
"Find 11 Owls"

A Parliament of Owls

Although this book focuses on attracting screech owls to your property, we recognize that not everyone lives where these species are the likeliest choice. Consider the following as an equal-opportunity owl guide based on geography and habitat.

Many of the criteria that make a screech box more attractive to prospective tenants may apply to these other species of owls, and the dimensions of the box or nesting structure can be adjusted based on the type of owl most likely to nest near you. The ways to prevent squirrels and other competitors/predators (see "Squirrels & Other Perils") may apply as well.

Even if most of these owls won't nest in your area, they're still worth learning about. With some planning and luck, you might find one on your travels. The owl species are listed from those with the largest populations to those with the smallest, with totals courtesy of Partners in Flight.

Can you find the eleven owls in this illustration?
(answers on page 227)

Owl-Struck

If you put up a nest box and an owl moves in, you may become owl-struck. Think of this emotional condition as a cross between awestruck and moonstruck, a mildly irrational attraction to all things owl-related.

For folks who have been owl-struck, Reddit even has a subreddit "dedicated to everything that makes you go 'Hoo'! Whether it's owl photographs, owl art, owl clothing, or asking how to help an owl you've found in need of help."

You know you've been owl-struck when you start buying every owl-related item you see.

Great Horned Owl

• •

US/Canada population estimate 3.8 million

Population trend Declining

Range Most of Canada, all of the mainland United States, and most of Mexico into Central and South America

Other common names hoot owl, feathered tiger, tiger of the sky

Vocalizations When most people think of owl calls, they think of the legendary *hoo, hoo, hoo* of the Great Horned Owl, but the call is more of a *hoo, hu-HOO, hoo, hoo.* Breeding pairs may alternate these calls, with the female's voice in a higher pitch.

• •

This species is the second-largest US owl (up to 5.5 pounds), just a few ounces behind the Snowy Owl. When the public thinks of owls, they tend to picture the majestic Great Horned Owl. Its large ear-like tufts resemble horns, which inspired its name. Its face is cryptically colored with shades of brown, rusty black, and white. The Great Horned Owl's wizened look has inspired many a logo, including the one for Wise potato chips. But the nickname feathered tiger is spot-on—Great Horned Owls have been filmed attacking nests of eagles three or four times their size.

Great Horned Owls are large powerful birds that nest early in the year.

Its piercing eyes are yellow, its throat white. Plumage varies depending on region and habitat. Some lighter subarctic Great Horned Owls might be mistaken for immature female Snowy Owls at a glance, while the plumage of their kin in the Pacific Northwest can be darker.

Great Horned Owls typically use large former nests of other raptors and herons. They also use cliff ledges and large hollows in larger trees, platforms built for Peregrine Falcons, and an occasional building ledge.

Before trying to attract Great Horned Owls to your property, consider that you'll need to install a sturdy basket the size of a large planter at least 15 feet up a stout tree and preferably much higher. So, hire a tree professional or attract a different raptor.

Barred Owl

· ·

US/Canada population estimate 3.5 million

Population trend Increasing

Range Eastern half of the United States and the Pacific Northwest; across southern Canada

Other common names hoot owl, striped owl, swamp owl

Vocalizations To communicate with freedom seekers on the Underground Railroad, nineteenth-century abolitionist Harriet Tubman used the Barred Owl's signature *who cooks for you, who cooks for y'all* call, which would blend in with the normal sounds of night. Barred Owls also have a raucous caterwauling duet that sounds like coyotes and monkeys throwing a party.

· ·

Primarily an owl of mature woods, forests, and swamps, this medium to large owl is smaller than a Great Horned Owl and larger than a Long-eared Owl. The Barred Owl has somewhat drab brown and white coloration, with namesake horizontal brown barring on the upper chest and vertical brown bars on a whitish belly. Its head is large and round, with large brown eyes and no tufts.

The Barred Owl population has benefited in areas where unused open space has returned to woodlands. Their numbers are growing in many parts of the United States as they expand their range northward and westward, but not without repercussions.

Barred Owls have become a predator of Western Screech Owls and a serious threat to Spotted Owls, the icon of environmentalists in the Pacific Northwest. Barred Owls tend to be more aggressive than Spotted Owls and can drive them off territories and even hybridize with them (these so-called Spotted × Barred Owls have also been dubbed sparred owls). Barred Owl habitat overlaps with Eastern and Western Screech Owls quite a bit. In the Northwest, where Barred Owls have recently become prevalent, screech owl populations tend to decline, likely because the screeches have yet to adapt to the newer predator.

Barred Owls also have adapted quite well to semi-urban and suburban environments with large old trees scattered about, especially in parts of the South. In

Barred Owl populations have increased drastically in areas where forests have overtaken farmland.

fact, the University of North Carolina–Charlotte conducted one of the most extensive studies of Barred Owls. Using nest boxes and tracking devices in and around Charlotte, researchers found that the Barred Owls had a nest-box occupancy rate of a healthy 30 percent.

Nest boxes for Barred Owls are necessarily large and heavy—usually 24 inches tall with a floor that's 14 inches square and a 7- to 8-inch-diameter entrance hole. The boxes are often installed by tree companies that have the equipment necessary to attach them at least 15 feet off the ground.

Northern Saw-whet Owl

. . . .

US/Canada population estimate 2 million

Population trend Increasing

Range In North America, southern Alaska across Canada, mountainous regions of the western United States and northwestern Mexico, and forested regions of northern and eastern United States; US and Canadian populations highly migratory in winter and may be found in all states

Other common names sparrow owl, farmland owl, little nightbird, little owl

Vocalizations The famed bird artist John James Audubon wrote that "the Little Owl is known in Massachusetts by the name of the Saw-whet, the sound of its love notes bearing a great resemblance to the noise produced by filing the teeth of a large saw." Not exactly. The repetitive metallic squeak from the axle of a rotating whetstone of yore was a dead ringer for the saw-whet *toot*. Perhaps owl expert Bernd Heinrich describes it best: "This tiny bird sounds like a city garbage truck backing up."

. . . .

The Northern Saw-whet is a small owl (7–8 inches tall and 2.5–3.5 ounces) with a large head, no ear-like tufts, yellow eyes, brown back with white spots, warm brown/rusty streaks down a whitish belly, and white and tawny-brown facial disks.

They breed in large tracts of northern, mountain, and temperate forest and swamps, particularly where evergreens are prevalent, especially pines. Finding

Saw-whet Owls are some-times mistakenly considered tame because of their reluc-tance to abandon their roost.

one of these migratory owls is a treat because they are so small and well camou-flaged, and they typically roost in dense evergreens. Sometimes locating a saw-whet requires luck, while other times scolding songbirds give away the owl's location. It usually takes downright persistence, checking every evergreen one comes across.

Saw-whet Owls are the ones most likely to be brought home accidentally in a fresh-cut Christmas tree. In 2020, a saw-whet made the national news when it took a 170-mile ride on a 75-foot-tall Norway Spruce destined for New York City's Rockefeller Center holiday display. The little guy was taken to a raptor rehabili-tation center, where it was released several weeks later none the worse for wear.

Saw-whets will nest in owl boxes, and a screech box with a 3-inch opening will work just fine. Ski towns in Colorado and Maine have established successful nearby saw-whet research centers and nest-box programs.

Boreal Owl

• •

US/Canada population estimate 1.5 million

Population trend Unknown

Range Subarctic and northern temperate regions of North America, Europe, and Asia; in North America, Alaska into Canada, northern Minnesota, and the mountains of the American West; rare nesting in the mountains of northern New England

Other common names Tengmalm's owl (in Europe)

Vocalizations These owls seldom vocalize outside of the breeding season. Their song tends to consist of whistled toots that get progressively louder. These toots are lower pitched and more rapid than the saw-whet's and sound quite similar to the winnowing sound produced by the tail feathers of a Wilson's Snipe during its flight display. (Take our word for it.)

• •

This is a small owl—about the size of a screech, a bit longer but typically a bit lighter, as it needs more feather insulation in its boreal habitats. Although most female owls tend to be larger than males, Boreal Owls take the cake: the female can be twice as heavy as the male. Its appearance is similar to a Saw-whet's but slightly larger, with a relatively large head. The facial disks are typically white around the eyes and grayer toward the outer border. The eyes are yellow, and the back is brown with white spots.

As their name suggests, these are primarily owls of the boreal forest, but they also inhabit mountainous subalpine evergreen forests within their range. Like the Saw-whet, the Boreal Owl is tough to locate, typically roosting in dense evergreens—not to mention that it typically stays far north. Boreal Owls are most often encountered in urban and suburban Canadian greenspaces in winter, as they leave the boreal forest during years of very successful breeding, severe winters, and scarce prey.

Typically, Boreal Owls are secondary users of woodpecker cavities, mainly those of Pileated Woodpeckers but sometimes flickers'. They readily take to Wood Duck

Boreal Owls are most visible in Canada in winter in natural areas near urban hubs.

boxes and screech boxes. There is growing interest in Alaska, Canada, and the US Rockies to use nest boxes to study the population of these remote northern forest-dwelling owls.

Burrowing Owl

. . . .

US/Canada population estimate 990,000

Population trend Declining

Range Southern central and western Canada, western half of the United States into Latin America and the Bahamas; quite migratory, especially US and Canadian populations

Other common names howdy bird, prairie dog owl

Vocalizations These owls make a variety of chatter and alarm sounds and can be quite vocal. Their song is a quail-like double note, *coo-cooah*, with an inflection toward the end.

. . . .

The Burrowing Owl is about the size of a screech owl but with longer legs. It has a relatively small round head with no ear-like tufts, white eyebrows, and yellow eyes. Its head and back are brown to sandy with white spots, and the belly is whitish with brown-sandy splotches and streaks.

Burrowing Owls live in open areas, grasslands, rangelands and deserts—even golf courses and airports. Nesting and sheltering underground, they thrive in areas with other burrowing animals such as prairie dogs, ground squirrels, badgers, and tortoises and will use other animals' abandoned burrows.

Like most small owls, they are opportunistic, with prey items that include insects, small birds, mammals, and reptiles. Burrowing Owls use their long legs to catch prey on the run but will also hunt from a perch or by gliding or hovering over larger insects. They are more active during daylight (diurnal) than many owls and will hunt any time of day or night, but they are most active at dawn and dusk (crepuscular).

Burrowing Owls are the stars of the book and movie *Hoot,* in which kids stop a developer from building on the owls' habitat.

Although Burrowing Owls don't use nest boxes, artificial burrow systems have been used successfully throughout their breeding habitat and range in the United States and Canada. First used in the 1970s and constructed mostly out of plywood, these systems now use plastic materials and involve a nesting chamber and about 6 to 10 feet of corrugated drainage pipe. The nest chamber is usually an upside-down five-gallon bucket, a heavy-duty storage bin, or half of a 55-gallon drum with a notch/hole for a drainage pipe. The good news is no precarious ladders are needed, just some sweat equity in digging a hole 3 to 4 feet deep, including the channel for the pipe that leads to an artificial burrow.

Burrowing Owls have adapted well to some suburbs and urban areas. The best example of this is Cape Coral, Florida, where an estimated 1000-plus pairs live inside the city limits. Manufactured burrowing systems can help sustain these populations.

Short-eared Owl

US/Canada population estimate 600,000

Population trend Declining

Range Widespread in the Northern Hemisphere throughout North America, Europe, Asia, and northeastern Africa; frequently, Northern Hemisphere populations are migratory or nomadic, breeding in the northern part of the range, wintering to the south, with year-round populations in the middle; also in South America

Other common names grass owl, marsh owl, prairie owl

Vocalizations Its song is a series of single-note hoots repeated somewhat rapidly, most often given by the male during flight display. Both sexes also produce a wing clap during flight display.

This species is similar in size to the Long-eared Owl, with sandy or tawny facial disks with yellow eyes surrounded by black. It has a round head with small ear-like tufts that are seldom seen. The back is cryptically colored brown, tan, tawny, black, and white. The belly is a dirty white; the upper chest has heavy tan streaking that transitions quickly to light to no streaking through lower belly.

Short-eared Owls are more active during the day than most owls. They hunt with buoyant moth-like flight over tracts of open land, using hunting grounds similar to those of Barn and Long-eared Owls.

When the Short-eared Owl's small tufts are flush while hunting, this owl can be tough to distinguish from a Long-eared Owl. Short-ears have also been known to occasionally join with long-ears in roosts, and they might be able to hybridize with them. This possibility was found when a graduate student at Montreal's McGill University was researching Short-eared Owls and found an unusual preserved specimen from 1990 in the collection of the Canadian Museum of Nature in Ottawa. The specimen's coloration and measurements were very consistent with a probable hybrid between these two types of owls.

Short-eared Owls hunt near dawn and dusk, flying low over large fields.

Unlike Barn and Long-eared Owls, Short-eared Owls nest on the ground, so unfortunately nest boxes or other structures won't encourage them to nest near you. They prefer undisturbed open lands so they can keep their nest concealed among the tall grasses or other low vegetation. They often nest on a mound and will shape a depression in the vegetation and even add some grass and feathers to the nest.

Long-eared Owl

US/Canada population estimate 150,000

Population trend Declining

Range Extensive in Northern Hemisphere, mostly subarctic to subtropical, across Europe, northwestern Africa, and Asia; in North America, Canada through the contiguous states; highly migratory, expanding range south and into Mexico primarily in winter

Other common names lesser horned owl, cat owl

Vocalizations Male has a single-note *hoo*, a bit ghostly, repeated with a couple of seconds between notes. The female version is similar but has a bit more vibration, often compared to a sound created with a comb and piece of paper. Long-eared Owls clap their wings as part of their display flights.

This medium-sized owl has orange facial disks with a black border and vertical black around the eyes. It has mostly brown streaks and barring on the chest and belly.

During the day it relies on camouflage, roosting in dense evergreen tangles or thickets. Finding a Long-eared Owl usually means accidentally flushing this seemingly secretive owl from its roost in thickets or dense cedar groves or finding whitewash and pellets under tall pine trees, then looking up. They roost up 50 feet or more in pines.

These owls typically hunt at night over open grassland, farmland, fields, or marshes (as do the Barn Owl and Short-eared Owl). Their namesake ear-like tufts are mostly visible when perched or roosting. Otherwise, they can be difficult to distinguish from Short-ears in low light. They are also somewhat nomadic and migratory. Alas, much of the US population of Long-eared Owls has declined as open lands are developed or transitioned back to woods and forests.

Known for their communal roosting, Long-ears form parliaments more than any other owl species in winter. Although most roosts attract a few owls, the world record for such a roost is the town square of Kikinda, Serbia. Some winters, more than 700 Long-eared Owls roost there.

Long-eared Owls frequently roost in dense evergreens near woodland edges.

They typically nest in former crow, magpie, or hawk nests but occasionally in owl boxes. The boxes we have seen in photos are from Hungary and more closely resemble nesting platforms with three walls and a roof, with a large front opening. Most conservation nesting projects worldwide use baskets 12 to 14 inches in diameter and 4 to 8 inches deep. Wicker lined with moss, leaves, feathers, or even coconut fibers seem to work quite well in suitable habitat. The basket is secured 15 feet or higher in the natural fork of a tree (preferably an evergreen).

Barn Owl

· ·

US/Canada population estimate 130,000

Population trend Increasing

Range Worldwide, mainly in temperate and tropical regions

Other common names ghost owl, silver owl, steeple owl, hobgoblin owl, screech owl (in Europe)

Vocalizations These owls screech or shriek at times.

· ·

This medium-sized owl has heart-shaped outline to whitish facial disks and no ear-like tufts. It is usually pale with whitish belly and underwings, buff to light tawny back and wings.

As its name suggests, the Barn Owl has adapted well to artificial structures and roosts in palm trees, evergreens or other foliage, barns, and silos. If you dissected an owl pellet as part of a class or workshop, most likely it was a Barn Owl pellet.

Like Long-eared and Short-eared Owls, they hunt over large tracts of open land, farms, ranches, vineyards, grasslands, and marshes with a buoyant flight. Barn Owls are excellent small-rodent hunters, and many farms and vineyards install boxes to attract them for natural and efficient pest control. In some areas, they have adapted to a more suburban lifestyle.

A pair of Barn Owls named Molly and McGee nested in an owl box with a nestcam in a California suburb. Their livestream had over 20 million views. Their popularity helped advance the residential use of Barn Owl boxes.

Nest-box designs are large and varied. The entrance hole should be about 6 inches in diameter and have a large floor area, at least 16 inches by 16 inches. Some look like mailboxes, and others like side-entrance buildings. Some are just enormous, deep birdhouses that are 24 inches from floor to the entrance hole. Some are made to be installed into the side of a building. In the right habitat, they all seem to work just fine.

Barn Owls, true to their name, frequent farm buildings and other human-made structures (including owl boxes).

Northern Hawk Owl

US/Canada population estimate 130,000

Population trend Unknown

Range Subarctic regions across North America, Europe, and Asia; in North America, Newfoundland to Alaska and in northern Minnesota, extending south into other northern parts of the United States during irruptive years

Other common names American hawk owl, Canadian owl, Hudsonian owl, day owl

Vocalizations A rapid high-pitched trill, like an Eastern Screech Owl's, but a bit higher pitched and sharper in tone. The trill can last up to 15 seconds, usually starts quieter and progresses louder. Also gives a rapidly repeated alarm call *kwit*, like an American Kestrel's or a Merlin's.

This medium-size owl has a roundish head, no ear-like tufts, and white-gray facial disks with a bold black border. The back is brown with white spots, breast and belly are whitish with distinct brown horizontal barring, and the tail is long. This owl does have a hawk-like appearance.

In the earliest known account of the Northern Hawk Owl (1831), Swainson, Kirby, and Richardson wrote that it was "more frequently killed than any other by hunters, which may be partly attributed to its boldness and its habit of flying about in the day."

Northern Hawk Owls inhabit more open areas of the boreal forest, frequently near bogs, marshes, or fire-damaged areas. They nest in natural hollows or old woodpecker cavities and occasionally in old nests of other birds. They have been documented nesting in open-fronted boxes with an 8-inch by 8-inch floor. The boxes were roughly 20 inches tall with a 4- to 5-inch board across the front to retain nesting material.

Northern Hawk Owls act more like hawks than owls, hunting during the day.

Northern Pygmy-Owl and Ferruginous Pygmy-Owl

• •

US/Canada population estimate 100,000

Population trend Stable

Range Northern Pygmy-Owl: forested mountainous regions of North America from British Columbia south to Central America, tending to move to lower elevations in winter; Ferruginous Pygmy-Owl: desert/mesquite areas of southern Texas and Arizona, coastal areas of Mexico into South America

Other common names mountain pygmy owl (Northern), cactus pygmy owl (Ferruginous)

Vocalizations A repeated high *toot* note similar to the Saw-whet Owl, but usually repeated at less regular and longer intervals. Also, a trill that is very rapidly repeated toots. Ferruginous vocalizations tend to be a bit higher in pitch than those of the Northern Pygmy-Owl.

• •

The Ferruginous Pygmy-Owl (pictured) and Northern Pygmy-Owl are known as fierce predators.

These tiny owls have a brown to rusty back with dark fake eye spots on the back of their heads. The facial disks are brown to rusty with white spotting, eyes are round and yellow, and the beak is yellowish. Ear-like tufts are present but seldom seen, and the belly is white with brown to rust streaking.

Northern Pygmy-Owls prefer forested mountainous regions within their range, whereas Ferruginous Pygmy-Owls prefer desert mesquite areas.

Despite their small size, pygmy-owls are feisty diurnal predators, known for preying on songbirds up to three times as heavy as they are. Both are nest-box candidates, as they nest primarily in old woodpecker cavities. Despite its limited US range, the Ferruginous Pygmy-Owl has better-documented nest-box use than the more prevalent Northern Pygmy-Owl. A scaled-down version of a screech owl nest box with a 2-inch circular opening would be fine.

Great Gray Owl

. . . .

US/Canada population estimate 58,000

Population trend Unknown

Range In North America, northern Minnesota and mountains of the Pacific Northwest south to Yosemite National Park; occasionally winters in northern New England; mostly non-migratory but will expand its range during irruptive years

Other common names phantom of the north, Lapland owl, striped owl

Vocalizations Song is a series of repeated low *hooo* notes that last 6 to 8 seconds, with a pause of 15 to 30 seconds in between. According to the legendary Arthur Cleveland Bent, "Several authors have referred to the voice of this owl as a tremulous, vibrating note, somewhat resembling that of the screech owl."

. . . .

If you saw the excellent 2011 birding movie "The Big Year," think back to that scene where Jack Black's character and his father come across a large and magical owl perched in a tree. *That* was a Great Gray Owl.

North America's tallest owl also has the largest wingspan, but at a maximum weight of 3.75 pounds, it is significantly lighter than a Snowy or Great Horned Owl because of its thick layers of insulating feathers. The Great Gray Owl has a large round head, yellow eyes, and no ear-like tufts. The plumage is white with light and dark gray streaking and many tones of gray. It has a distinctive white bowtie mark below the beak.

Great Gray Owls prefer areas of the boreal forest with scattered openings of meadows and marshes. They generally nest in the former nests of raptors or ravens, in large hollows, or in chimney-type depressions within the broken trunk of large-diameter trees. Nesting platforms similar in size to Great Horned Owl baskets or platforms have been installed 15 to 30 feet up in spruce or fir trees within their range.

Great Gray Owls' feather insulation makes them look large, but they weigh less than Snowy or Great Horned Owls.

A. Williams · *Owls* · 93

Elf Owl

• •

US/Canada population estimate 40,000

Population trend Thought to be declining

Range Across Mexico and in the southern desert regions of Arizona, New Mexico, and Texas during breeding season; migrates to Mexico for winter

Other common names dwarf owl

Vocalizations A range of repeated high *whi-whi-whi* or *chur-chur-chur* notes.

• •

As its name implies, the Elf Owl is the smallest owl in the world, up to 5.5 inches and weighing only 1.9 ounces. It has a round head, white eyebrows, no ear tufts, and yellow eyes. The plumage is various mottled shades of brown and gray with rusty highlights and white spots, with dense streaking on the chest grading to a white lower belly and under tail.

A nocturnal insectivore famous for nesting in woodpecker cavities in saguaro cacti, this owl also nests in old woodpecker cavities in trees and telephone poles of desert and hardwood riparian areas in their range.

Elf Owls will use nest boxes. Provide them with a scaled-down screech box with a 2-inch opening.

Snowy Owl

• •

US/Canada population estimate 15,000

Population trend Declining

Range Breeds in Arctic regions of the Northern Hemisphere; migratory and irruptive (prone to sudden population shifts); winter range extends much further south

Other common names white owl, polar owl, Arctic owl

Vocalizations Song is a very low, rough, powerful single-note *hoot*, usually two at a time, with a brief pause between sequences.

• •

This is North America's largest owl by weight, with a maximum of 6.5 pounds. It has a round head, yellow eyes, and small ear-like tufts that are seldom visible. The plumage is mostly white, with varying amounts of black markings. Mature males have the least number of black markings, whereas immature females tend to have the most.

They winter in open areas such as farm fields, airports, landfills, beaches, and marshes. Due to their daytime activity, large size, charismatic looks, and propensity to perch or roost on conspicuous places (streetlights, roofs, jetties, and tops of dunes) on their wintering grounds, Snowy Owls are one of wildlife lovers' favorite owls to seek in southern Canada and the northern United States.

Good nesting success or lack of accessible food can drive more Snowy Owls south for the winter. During the winter of 2013–2014, for example, there was a super

Snowy Owls are large, conspicuous, and charismatic, which makes them a treat to see.

irruption of Snowy Owls. This was foreshadowed when researchers found a nest surrounded with a cache of seventy lemmings and eight voles in northern Quebec the previous summer. This prey abundance on their breeding ground led to a very successful breeding year, and that winter at least one Snowy Owl made it to Florida. Although their population still seems to be in decline long term, the interest generated during that winter helped fund quite a bit of continuing research to get a better insight into Snowy Owls on their Arctic breeding grounds.

More recently, a Snowy Owl made headlines by flying around Washington, DC, and making stops at the National Postal Museum, US Senate buildings, Capitol Police headquarters, and the statue of Archimedes above the entrance to Union Station.

Because Snowy Owls nest on the tundra in the Arctic, artificial suburban nesting structures are superfluous, so no need to rush out to look for Snowy Owl nest-box plans.

Spotted Owl

· ·

US/Canada population estimate 15,000

Population trend Declining

Range Western North America from southern British Columbia to central Mexico.

Other common names canyon owl, brown-eyed owl, wood owl

Vocalizations Typical song four- or five-note hoots, somewhat similar in tone to a Barred Owl song. They also produce a combination of hoots and whiny barks.

· ·

This is a medium to large owl, a bit smaller than Barred Owl. Its plumage is dark warm brown overall, with some tan highlights, with white spots on the back and broad oblong white spots on the belly. It has a round head, no ear-like tufts, and dark eyes.

The Spotted Owl has been the icon of Pacific Northwest forest conservation, even making the cover of *Time* in 1990. The species is threatened not only by the loss of old-growth forests, it's also being displaced by the Barred Owl.

Spotted Owls typically nest in old raptor or corvid nests, chimney-type depressions in broken trunks, or large natural hollows. A Barred Owl box or half of a whiskey barrel installed 15 to 30 feet up a tree should work.

Spotted Owls are the poster owl for efforts to conserve Northwest old-growth forests.

Flammulated Owl

· ·

US/Canada population estimate 11,000

Population trend Unknown

Range Highly migratory; breeding range in Mexico and western half of the United States to British Columbia; winters in Mexico to Central America

Other common names flammulated screech owl

Vocalizations Deceptively low-pitched hoots suggestive of a larger owl. Flammulated Owls have a reputation for throwing their voices like a ventriloquist, making them difficult to locate even if nearby.

Flammulated Owls resemble screeches but have brown eyes.

· ·

This is a tiny owl, smaller than a Saw-whet. The Flammulated Owl resembles a tiny gray-morph screech owl with much shorter ear-like tufts, large brown eyes, and with rusty highlights around the face, eyes, and back.

The Flammulated Owl is primarily a nocturnal insectivore. Despite the species' long-distance migrations, they tend to have strong fidelity to breeding territory and even nest site, returning to the same nesting area if successful. They nest in mountainous mixed forests with a prevalence of ponderosa pines or other mature evergreens.

Flammulated Owls nest primarily in old woodpecker cavities but also use nest boxes. A slightly scaled-down version of a screech box with a 2.5-inch circular opening has worked well.

Hosting a Screech Owl B&B

Enticing screech owls to roost and nest near you can be magical, especially with a miniature video camera to record the action inside the box. Most people go through life without seeing an owl in the wild, let alone having one for a neighbor.

The guides on the whale-watching boats out of Gloucester, Massachusetts, start every trip by informing the passengers that they aren't going to SeaWorld—they are hoping to view wild animals in their natural habitat. They could end up with a stellar experience, but there were no guarantees.

We offer a similar caveat for putting up an owl box for screeches. They are wild, free to come and go as they please. Natural and human-induced risks to a small owl are always there, ranging from predators and starvation to poisoning and death by automobile. On the other hand, to paraphrase hockey legend Wayne Gretzky, screech owls miss 100 percent of the boxes that aren't up.

An Eastern Screech Owl is tucked snugly into an owl box.

We have four decades of combined experience and observations of all things screech owl—especially nest boxes, site selection, and installation. Taking our advice won't guarantee getting screeches for tenants, but your chances should certainly improve.

Screech owls are very adaptive to urban and suburban areas with a few large trees and green spaces, as well as woods, orchards, and swamps. Unlike Red-shouldered Hawks and some other raptors that select new nesting sites every few years, once screeches choose a nest site, they tend to return. And even if they don't, you stand an excellent chance of attracting new tenants.

Attracting a screech to move into your owl box is akin to attracting a homebuyer. Ideally, you want the home to be move-in ready and in a great location. You also want to work on the owl equivalent of curb appeal—making the box and its surroundings irresistible to an owl pair looking to raise youngsters.

And like real estate, the law of supply and demand applies. Much of an owl box's success depends on how many screech owls are in the area and how many other nesting cavities are available. Screech owls seem to look into increasing their cavity portfolio year-round. Even if it's just a brief midsummer, midnight visit, they investigate and inventory new and existing cavities. Whether the result of competition, damage, or new construction, cavity availability is always in flux.

Attracting an Owl

The first question is whether screech owls live near you. If you've heard or seen screech owls in your area, you're in business. If you haven't, the best place to start is eBird. This global citizen-science project provides real-time and historical sightings, including species maps, for almost every bird species in the United States and Canada.

The eBird app is both a curse and a blessing for people who love owls. It can be a curse because if too many people find out about an owl's location, they may unintentionally harass it in search of a good photo or a so-called life bird—a species that a birder has identified in the wild for the first time. But eBird can also tell you about prospective resident owls. Go to ebird.org/map, enter "Eastern Screech Owl" or "Western Screech Owl" for the species name (depending on where you live), and enter your location. If the map has any blue markers (sightings older than 30 days) or orange markers (within 30 days) in the area, you're probably good to go.

If the answer is "no," you may want to target a different type of owl. If you live by deep woods, like in central Maine, you should investigate Barred Owl boxes or

Great Horned Owl nesting baskets. You may want a Barn Owl box if you live by extensive farmland, large vineyards, or other open areas.

One other caveat, courtesy of Robyn Bailey, project leader for the Cornell Lab of Ornithology's NestWatch citizen-science program: If you know of a nearby Barred Owl or Great Horned Owl nest, think twice about putting up a screech box. When nesting, the smaller owls are easy targets for these larger predators.

Similarly, if a larger owl establishes its territory near a successful screech nest box, this can lead to the displacement of the resident screech. Such is the nature of survival for a smaller owl.

Choosing the Right Spot

Once you've determined your property is ripe for screech owls, it's time to choose the best place for the owl box. First, find a suitable tree or locale for a post or pole, away from utility lines.

One question we're most frequently asked is what direction the owl box should face. Based on extensive experience, we advise any direction but north. If you live in a colder climate, research suggests that screeches may prefer entrances that face toward the southern half of the compass, which will help the owls get a bit of warming sun, especially after the leaves have fallen. If you live in a warmer climate, make sure the box receives adequate shade during the heat of the day.

Next, consider the habitat, mindful that screech owls tend to have a U-shaped flight path—they drop from their perch, fly lower, and fly up to their next perch. An area with mature trees, canopy cover, and little or an uncluttered understory is ideal. Many suburban yards, old cemeteries, and parks fit this category. Solitary trees in or adjacent to a field are more suitable for American Kestrels.

Screech owls have excellent eyesight, but they need to see the entrance to find a potential nest site. A building or dense greenery that obscures the nest box's entrance may prevent a passing screech owl from seeing the potential dwelling.

The flatter the terrain, the safer the installation. Hills make installing a nest box difficult and unsteady. Facing a box downhill gives the impression of a higher box, while facing it uphill gives the opposite impression.

The ability to view the owl box during nesting season from the comforts of your home is a major reason to erect a box in the first place. It's an unobtrusive way to enjoy owls—and to keep an eye out for squirrels and other interlopers. Just make sure you have taken steps to make your windows bird-friendly (see "Owl Conservation Issues and Ways to Help").

This screech box is a mixed bag. It's more than 10 feet up on a straight tree trunk, but it's in the open and facing north.

Proximity to your residence is not a big concern. In fact, screech owls have used boxes attached directly to buildings. Just keep in mind that the closer the box is to your home, the more the box should face the building at an angle.

Another factor is how much activity that part of your yard gets. Choosing a quiet spot is more important than the nest box's proximity to your residence. Screech owls rely on their hearing and make a lot of barely audible sounds around the nest, so you should keep the nest box away from constant noise sources. For instance, in the South air-conditioning units may cause a problem; they are seldom a problem in northern climates because these units are not used regularly before early June, after the screech owl nesting season typically ends. Sporadic noise such as weekly lawn mowing does not seem to dissuade screeches from using a nest box, but louder noises such as chain-sawing trees near an occupied box may keep owls away for a while. Remember, each owl is different and will have different responses to disturbances.

A third factor is whether you are installing a nestcam. At this writing, the best cameras are wired for electricity, image, and sound. The farther you get from your residence, the more wiring it entails and the greater the chance of connectivity problems.

Above: This Pennsylvania kestrel box attracted a roosting Eastern Screech Owl even though it was located by a busy road and a field.

Above right: An Eastern Screech Owl roosted in this kestrel box for an entire spring.

Box Placement FAQ

What about automobile traffic?

Roadside roosts for sunning are common in the winter, but owls typically do not use these locations to nest. Traffic and road noise are not ideal for nesting, and the screech owls' U-shaped flight pattern makes them susceptible to vehicle strikes. Although not ideal, near a driveway shouldn't be too much of an issue; install the box as high as is comfortably possible.

What about kids, dogs, outdoor cats, deer?

Children playing in the yard do not seem to affect screech owl box habitation too much, as long they aren't bothering the tree or post that the box is on. With active kids, dogs, or cats, it may be prudent to install the box on the higher end, say 15 to 20 feet, giving the owl more distance from the activity.

Dogs and cats are primarily a threat to owlets, and a parent owl will attack if the pet is perceived to be a threat, especially just after fledging. Squirrel prevention measures (see "Squirrels & Other Perils") also apply to cats. How many active dogs are in the yard will likely affect screech residency. Owls seem to be unfazed by deer, since deer are herbivores. We have seen deer graze under a screech owl sunning itself, and the owl couldn't have cared less.

What about placing screech boxes in nature preserves?

It sounds like a match made in heaven. But based on our experiences—both positive and negative—it can be more complicated.

Unless the owl box is in a location that people can't access, we offer these warnings. Any time a nest site receives too much attention, the birds may abandon the location altogether, including the eggs. In general, the more time and effort that have gone into nesting, the more stress it takes for the owls to abandon the site. Of course, minimal attention from a distance is best.

If a nest site is discovered after the eggs hatch, the parents may not return to the nest with food if a human is nearby. Trying to feed all those young mouths is stressful enough—don't make matters worse.

When people get close to a nest, they increase the chances that raccoons, foxes, and other predators will find it. Wherever humans go, they leave a scent trail, and predators may follow that scent in search of a meal.

Getting too close to an active screech box may increase the chances that raccoons and other predators will find it.

If you insist on placing a nest box on public land (with permission, of course!), put it off the beaten path and as high off the ground as possible—a minimum of 16 feet. Try to keep sightings off eBird, online forums, and social media until well after nesting season, if ever.

One Owl Box or Two?

The natural tendency is to put up one box. After all, unless you're sure you have screech owls in the area, why go to the expense and trouble of putting up two?

Owl aficionados who've installed two boxes have found that both are used regularly during nesting season and occasionally other times of the year. The male sometimes roosts in the second box and uses it as a pantry to store prey, especially when a major event is about to happen—when the eggs hatch or when the owlets start to fledge, for example.

A second box adds another dimension to the experience. The key is choosing good locations for both boxes so that you're all set if you do get nesting screech owls. Placing the boxes near each other (say 10 to 100 feet apart) allows the male and female to stay close and allows the male to watch the other nest box. Two boxes also allow you to confirm you have a pair of screeches if each entrance has an owl in it at the same time. Typically, unpaired owls will not roost near each other and will try to defend the prime real estate from intruding owls.

Also keep in mind that if you install a camera in only one of the owl boxes, place the other box roughly 8 feet off the ground so it's not as attractive a place to nest. You want to watch the action inside the nest box, not the pantry.

Screech territories vary widely but are roughly 0.5 mile by 0.5 mile (0.25 square miles). If someone else has a screech box within a quarter mile of yours, you may share your owls. Boxes (paired or unpaired) within a quarter mile will tend to share owls unless there's an abundant owl population. That means an owl pair's territory would include both boxes or both sets of boxes, and the owls will use them to roost at their convenience and may choose one box to nest. It may be yours, or it may be the neighbor's.

Never Take Hosting an Owl for Granted

One of the many underappreciated things that we love about screech owls is the relative ease with which they might be induced to occupy a nearby nest box in suburban and even urban residential areas, especially when compared to other owls.

Most owls wouldn't think of moving into a nest box near humans, and even those that might—Barn Owls and Barred Owls—are long shots.

Take Jim's longtime friend and mentor, Stiles Thomas. Back in 2014, he was in his early nineties and still living nearby, and he called to say that he had just heard a Barred Owl calling near his back porch. He wondered if I could come over and bring my camera.

A chance to see a Barred Owl in the wild? This charismatic owl has been on New Jersey's threatened and endangered species list since 1979. I wouldn't have missed it for the world.

As soon as I arrived, Stiles and I walked softly around the large backyard and surveyed the trees. You'd think we would have no trouble finding a large brownish white owl, but you'd be wrong.

With nightfall descending, we rescanned the limbs of every conifer behind the house. That's where the owl was most likely to perch.

Nothing.

After 10 minutes of futility, Stiles suggested we call it quits.

I said that before we gave up, Stiles should try his throaty Barred Owl call, *Who cooks for you? Who cooks for you all?*

Stiles demurred. "I'm too old for that to work anymore."

I cajoled and insisted.

Finally, Stiles muttered, "Alright, alright," and did his Barred Owl imitation. Then we waited.

Not a peep.

"See, I told you," Stiles said finally.

At that moment, I happened to look up. What did we see on a tree branch but a Barred Owl staring down at us with those soulful coal-black eyes.

We gasped. I slowly raised his camera and took a photo in the fading light. Then Stiles and I retreated to his back porch.

After our hearts stopped racing, we couldn't stop talking about what we'd just witnessed. Then Stiles soon put the whole encounter in perspective. "You know, back in the 1950s or so, we had a field back there," he said, pointing to a stand of evergreens. "Then one day I thought that if I grew a pine grove there, someday I'd get a Barred Owl."

Many folks can't plan more than a month or a year out. Stiles had been thinking six decades ahead.

Soon after, a friend built a huge nestbox for Barred Owls to place in Stiles' backyard—it was big enough to fit a Saint Bernard. They enlisted a tree service company to put the nest box on a maple trunk, but no Barred Owl ever moved in. Several years later, another company relocated the owl box to a local nature preserve. It has yet to attract a tenant.

Screech Owl

All night each reedy whinny

from a bird no bigger than a heart

flies out of a tall black pine

and, in a breath, is taken away

by the stars. Yet, with small hope

from the center of darkness

it calls out again and again.

—Ted Kooser, *Delights & Shadows*

An illustration by the artist Tony Angell, from his book *Owls*.

Squirrels and Other Perils

If you find it odd to read a chapter about squirrels, consider yourself blessed. In a perfect world, you'd attach an owl box to a tree on the edge of your backyard, wait for nesting season, and then sit back and enjoy the show. Alas, nature and (all too often) humans have a way of interfering. Over the years, we've found that squirrels are owl box Enemy No. 1. These bushy-tailed busybodies consider the boxes a great place to find shelter in the winter and raise families other times of the year—if you let them.

In all our years of writing about screeches and dealing with owl boxes, one question never goes away: "What can I do about the $%^!@#% squirrels?" Yes, some folks love squirrels. They find their persistence and mischief appealing. If you're one of these people, this is not the chapter for you.

Squirrels are owl box Enemy No. 1.

When squirrels commandeer an owl box, screeches will nest elsewhere.

Entire books have been written about thwarting squirrels. With titles like *Outwitting Squirrels* and *Squirrel Wars*, they appeal to everyone's inner Elmer Fudd. At last count, Mark Rober's video of a squirrel outwitting an obstacle-course maze had more than 68 million views on YouTube.

Lest you think we have nothing good to say about squirrels, however, consider this. Like most things in life, squirrels are good in moderation, and they inadvertently boost the oak population because they hide acorns all over. Unlike a Tufted Titmouse, which can remember where it buried thousands of seeds, squirrels have lousy memories. Mighty oaks grow from these forgotten acorns.

Chipmunks don't pose a threat to owl boxes, because they live in burrows. You likely won't find one in an owl box unless it's a meal.

The Trouble with Squirrels

Five types of squirrels compete with Eastern and Western Screech Owls for housing: the eastern gray squirrel, fox squirrel, pine squirrel, flying squirrel, and western gray squirrel. By far the two most troublesome are the eastern gray squirrel and the fox squirrel. Both weigh much more than a screech owl and win most battles.

Pine squirrels, which include the widespread American red squirrel, are equally feisty. They weigh about as much as a screech, but the owl tends to prey on them or outcompete them for a box. The flying squirrel also uses owl boxes, but it's easy pickings for a screech. The western gray squirrel, native to the West Coast, resembles the eastern gray and size-wise is between the eastern gray and fox squirrel. But western gray squirrels are far more timid and tend to be averse to humans and suburbanization.

Any owl conflict with a large squirrel can lead to injury or death, even if the squirrel takes a beating as well. The no-holds-barred exception is when screech owls defend an active nest—then it's winner take all. And to add insult to injury, squirrels don't need to use tree cavities. They're quite capable of building their own shelters (or dreys) with sticks and leaves. People often mistake these shelters for bird's nests.

Female fox and gray squirrels do relocate nesting dens before their pups are fully weaned, typically in response to a disturbance or flea infestation. Protective momma squirrels will do whatever it takes to keep her pups safe and dry. In one instance in Texas, a wet fox squirrel in search of a dry shelter invaded a nest box containing a female screech and her three nestlings. The squirrel killed two owlets in the scuffle and came back the following day and killed the third. The screech wisely relocated and started a family in another owl box.

The Texan who witnessed the fracas via a nestcam became obsessed with evicting squirrels after that. He devised owl boxes that included built-in deterrents with names like The Hammer of Doom (involving a giant mallet) and The Fingers of Death (involving wooden rods that arose from the floor of the box). Despite the colorful names, the devices were more Rube Goldberg than Torquemada.

You needn't go to such extremes. Yes, fox and eastern gray squirrels are tough and wily adversaries. And, yes, you'll probably never get the better of them. But with a little work and investment, you can reduce the aggravation.

The squirrel question often arises in January or February. Should you risk opening the owl box to see if squirrels have moved in, or should you leave well enough alone in hopes a screech will drive out the interloper? The long-term answer is to switch to a squirrel-resistant box (see "Building and Installing a Squirrel-Resistant Screech Box") or implement better preventive measures in time for the next nesting season.

The short-term answer, however, is that if you see a squirrel in the box, evict it immediately and remove any leaves or other bedding material it has imported. Otherwise, screech owls will look somewhere else for a place to nest.

North American squirrels come in five basic varieties, including (clockwise from top left) fox squirrel, red squirrel, flying squirrel, and eastern gray squirrel (this one's melanistic or black).

Sometimes the sound of putting a ladder against the tree is enough to make the squirrel scram. Sometimes a bang on the side of the box will do. But other times a squirrel will hunker down, and occasionally more than one squirrel will lurk inside.

If you think the squirrel is gone, you can be in for a dangerous surprise when one squirrel and then another leap out while you're standing on a ladder and preparing to clean out the box. When a squirrel switches to a flight response, it can be explosive. It's best to approach the opening at an angle to allow a clear escape route for the squirrel. If a squirrel can survive a leap from the twenty-first floor of a building—and they have—then 15 feet from a nest box is nothing.

To make certain, slowly open the box and use a child's rake or small gardening tool with a 3- to 4-foot handle to jostle the bedding/leaves. Once all is clear, remove the leaves. In general, the easier it is to evict the squirrel, the quicker it will return at first. The harder it is, the longer the squirrel will stay away. Repeat as necessary. Screech owls will roost in nest boxes partially filled with leaves, but they won't remove the leaves themselves and will likely nest elsewhere.

If you don't see a squirrel entering or leaving the box, watch the opening at dawn and dusk. If you see an owl, leave well enough alone or—if you strongly suspect lots of leaves have been imported—do a quick check of the box at night if you are comfortable on a ladder.

You can also use a telescoping mirror or a small cellphone camera on a selfie stick to check the box contents. Some cellphones may be too wide for a 3-inch-circular opening, but smaller and older cellphones should fit. If you have access to a GoPro or even a small nanny-cam, use that.

Squirrels overran this Washington, DC, park after humans provided them with food and nest boxes.

The Squirrels of Lafayette Square

A classic squirrel infestation occurred more than a century ago in Lafayette Square in Washington, DC, where researchers tallied what is thought to be the highest squirrel density ever recorded. The 7-acre park had as many as 200 squirrels living there.

The problem likely began in 1907, when a group called the Fish and Game Protective Association released a variety of birds and mammals in the city's parks. Included in the mix: "Lafayette Park, 48 gray squirrels."

Five years later, *The Washington Post* reported that park officers in Lafayette Square were feeding the squirrels a bushel of nuts a month. They also erected fourteen little houses for the squirrels to live in. Visitors began feeding the squirrels as well—roughly 75 pounds a week. So many squirrels overran the park that they also devoured tree bark and plantings. According to *The Post*, the squirrels once ate more than $2200 worth of geraniums in two days.

Authorities were compelled to intervene. They tried squirrel birth control, relocation, and ultimately removing the nest boxes and outlawing the feeding of squirrels. The furry-tailed rodent population eventually dropped to a more manageable level.

Moral of the story: If a lot of squirrels are hanging around your yard, especially in late winter and early spring, you need to make your feeders more squirrel-resistant and your bird buffet less appetizing. Ideally, you'll need to enlist the aid of any neighbors that have bird feeders as well.

Keeping Squirrels Away

Once you have evicted any squirrels and removed the leaves, take as many preventive measures as you can. If the owl box sits atop a post or pole, for example, you should put a baffle on the post to make it more difficult for a squirrel to gain access. The base of the baffle should be at least 3 feet off the ground. If the baffle isn't placed high enough, squirrels have been known to latch onto the base of the baffle with their teeth and gradually use their body weight to lower it enough for them to overcome it. With squirrels, expect the worst.

If the nest box is on a tree, wrap an 18-inch-wide strip of sheet metal around the base of the tree, about 3 feet up. If possible, add another strip of flashing above the box. That's about as squirrel-proof as a typical owl box can get. The flashing may not stop all squirrels since they are expert tree climbers, but it will make access more difficult for other predators, including house cats, snakes, and raccoons. You can secure the flashing with an oversized hose clamp or two.

Backyard squirrel populations tend to expand with the amount of food available. Thus, if a lot of squirrels are freeloading at your bird feeders, sooner or later they will try to commandeer your owl box(es).

Aside from thistle feeders and suet feeders placed inside simple metal cages, which you can buy at a birding supply store or online, most squirrel-resistant feeders are ingenious if a tad complicated. They typically come in two varieties: gravity regulated and motorized, with key parts made of metal to prevent squirrels from gnawing their way to the seed.

With the gravity-regulated feeder, the squirrel's own weight on the feeder's perch shuts off access to the seed. Motorized feeders have beguiling names like Twirl-a-Squirrel, Fortress, and Yankee Flipper. They operate with a motor that spins when a squirrel lands on the feeder's perch but won't activate when woodpeckers and other large (but lighter) birds land there. Although the motor rotates at speeds that will eject the squirrel without harming it, these novelty devices are not everyone's cup of tea. Why amuse yourself at another creature's expense?

A somewhat controversial deterrent is spicy birdseed, which comes with names like Fiery Feast or Wild Delight Sizzle N' Heat. The seed's cayenne pepper or capsaicin coating makes a squirrel's mucus membrane smart. (Oil of peppermint is said to be a deterrent as well.) Birds, meanwhile, can't taste the pepper, but the coating may affect their eyes in certain situations—so less is more. The spicy seed has the added benefit of keeping the local white-tailed deer population from freeloading at squirrel-proof feeders.

Although these pepper-laced products cost more than regular birdseed, you can mix six parts regular seed with one part hot seed, and the squirrels will still get the hint. And since squirrels typically eat their body weight in birdseed and other food each week, you'll save money and trips to the bird-supply store in the long run.

In case you're wondering if squirrels hate peppermint and hot peppers, why not put some in the owl box? The answer, according to Steven Sullivan of Project Squirrel, a citizen-science group that collects gray and fox squirrel observations from people around the world, is simple: "I'd be hesitant to use them in an enclosed box nest because the scents can be really irritating to the owls' and owlets' tender skin and eyes."

What We *Don't* Recommend

Clever humans have devised other ways to thwart squirrels, but some strike us as inhumane and others come with unintended consequences.

Shock treatment

Squirrels may be a screech-box nemesis, but that's no excuse to be cruel. A few books that offer tips on squirrel-proofing suggest buying feeders that include a mild electrical shock that you can deploy with a battery-powered remote control. Not only is it complicated, but you must watch your feeder in order to zap the squirrels. Although one manufacturer claims "the shock isn't strong enough to fry the pesky little tree rats," it still sounds nasty.

Snap traps

Killing squirrels just seems way too extreme, especially when squirrel-resistant owl boxes and other deterrents are available.

Relocation

Several traps allow you to catch the offending mammal so you can relocate it instead of killing it. This may seem more humane, but the squirrel will soon find its way home if you don't take it several miles away. If you do successfully relocate it, you are placing it in a new habitat with an existing squirrel population that probably won't greet it with open arms (or whatever squirrels call their front legs). In the winter, you'll also cut the squirrel off from its caches of nuts. No wonder it is illegal in many jurisdictions to relocate wild animals. Sullivan calls relocating squirrels "unethical and ineffective."

Other Nest-Box Invaders

Other formidable screech owl nest raiders include skunks, weasels, raccoons, opossums, ringtails (a southern raccoon relative), and fishers (a northern weasel relative). Owl boxes with 3-inch-diameter openings are more effective in staving off many of these predators, and boxes with larger holes or slots should be deeper and narrower to prevent these mammals from reaching in and grabbing an owlet.

The good news is that anything that makes it difficult for squirrels to take over a box will make it almost impossible for other mammalian predators. Cats, foxes, and coyotes not only will prey upon screech owls, but they'll also compete for the same rodents and birds.

Avian predators of screech owls are numerous as well. Cooper's Hawks seem the most likely to find a roosting owl, but most larger raptors will prey upon them if the opportunity arises. At night, any owl larger than the screech can pose a danger. Although we have seen screech owls successfully contend with a couple of Snowy Owls, wintering Barred Owls, and (for a few days) a Great Horned Owl within and/or next to their territory, the frequent presence of a larger raptor such as a Red-shouldered Hawk near the nest box may prompt a screech owl pair to relocate at the beginning of the breeding season.

Starlings can be pests as well. The European Starling is considered an invasive species that has had a detrimental effect on many native cavity-nesting birds, such as kestrels, flickers, and bluebirds. Although they may be on the menu for screeches, starlings are aggressive (with a long sharp beak) and persistent enough when competing for a nesting cavity to occasionally get a screech owl to abandon its nest. They also tend to empty out a box of all the bedding material before building their own nest.

A Red-shouldered Hawk perches on an owl box in late winter. The screech owls roosting there moved on.

Reptiles are an uncommon threat. For the most part, smaller lizards and snakes are meals, but some larger snakes will raid a screech owl nest. Rat snakes are widespread constrictors and known nest raiders. Green iguanas are large enough and can climb trees and take eggs and recently hatched nestlings, but they are currently limited to southern Texas and southern Florida.

Insects are also pests to be aware of. Annually cleaning the nest box out and replacing wood chips after the owlets have fledged is the best way to prevent many insects from becoming firmly established. But minimizing surprises on a ladder is always a good idea as well.

Bees, hornets, and wasps can all build hives in an owl box, and this can prohibit owls from using the box if hives become established and present a danger to owlets. In warmer climates, it's not unusual for screech-box owners to take the box up or down based upon seasonal bee swarming activity and screech owl nesting activity. Wasps and hornets can also be a problem when cleaning out a box after nesting season. Just observe the opening for a bit before opening the box to see if any stinging insects are coming and going.

A light coating of bar soap or cooking spray inside the roof is a frequently suggested means of preventing bee, hornet, and wasp hives from getting started. Their effectiveness wears off over time, so the trick is remembering to reapply when cleaning out the box.

In case a hive does get started, in regions that get below freezing regularly, it's easy to deal with insects in winter, when the insects go dormant. Just make sure no owl is present and disturbed when removing the hive. In areas that are warmer year-round, it's best to call professionals. In all regions, if the very beneficial honeybees have taken over the box, a local beekeeper can likely relocate and use them. The entire box with the hive inside may need to be removed, and the box may not be salvageable, but the environmental benefits are worth it.

Carpenter bees like weathered wood to make tunnels for their eggs, but they are solitary and don't do much damage on their own. The trouble arises when a woodpecker finds tasty bee larvae in the wood and starts excavating. To thwart carpenter bees, which look like bumblebees but have a shiny abdomen, we recommend using white pine for your screech boxes and painting the end grain on the roof and bottom.

Ants are also a possible pest, although there have been instances of beneficial symbiosis between screech owls and tree ants, such as the velvety tree ant. They've been known to reside in nest cavities and attack intruders but leave the owlets alone. Unless the infestation is severe, most ants can be evicted at clean-out time. The invasive fire ant is a concern in southern states and is currently expanding its range to the north and west, including locales in California and New Mexico. These ants bite and sting (similar to a bee or wasp sting), and they can and will attack an owl nest.

Finally, fleas and mites can be detrimental to the health of owls and owlets. They can and will build up in a nest box that is not cleaned annually.

We would caution about using the widely available pyrethrin/permethrin wasp and ant sprays to address insect infestations. While permethrin is reportedly not highly toxic to birds, nestlings may be more sensitive to the residue and its toxic effects. It can be nasty stuff for humans, too.

All About Screech Boxes

Screech boxes are designed to emulate natural tree cavities where the owls like to nest. Typically created by Northern Flickers or Pileated Woodpeckers, these cavities can also be hollows formed by limb rot and tree rot. Although there is no standard shape or size for cavities that screech owls will use, some cavities and entrances are too small or too large.

Even the male screech owl doesn't know which cavity his mate will choose for a nest. He just hopes one of the natural and/or man-made possibilities in their territory will be satisfactory to her. To help her choose, the male entices her to as many potential new homes in their territory with as much food as he can muster.

A red-morph Eastern Screech Owl sits at a nest-box opening.

Researchers have spent untold hours measuring natural-cavity screech owl nests to determine how wide the openings are, how deep below the openings the bottom of the cavities are, and the size of the floor. The result is the quintessential screech box: an average of all those measurements. Audubon.org and Cornell's NestWatch.org offer free directions to build a standard screech box, and in the next chapter we offer plans for a squirrel-resistant owl box designed by Scott.

Several other styles of screech boxes are on the market. Most are made from wood, usually cedar or pine. Cedar tends to be more expensive but it lasts longer, whereas pine is more prone to insect damage and decay.

Screech owls frequently use tree cavities with large entrances. With that in mind, Cliff Shackelford, a retired ornithologist with the Texas Parks and Wildlife Department, designed a box and has marketed it as the OwlShack since the early 1990s. It features a roughly 4.5 by 5.5 inch rectangular opening, and the floor (4.5 by 5.5 inches) is roughly 13 inches from the bottom of the entrance. The box's narrowness compensates for the larger opening to deter larger predators. This design has become so popular that many box builders and online how-to videos have imitated it.

Other screech boxes are made from (often recycled) plastic. These will basically last forever, but the aesthetics are not for everyone. Since the plastic is a bit slick, these nest boxes should provide more protection from squirrels and other predators than wooden boxes. If you're shopping for this type of box, also consider plastic houses made for Wood Ducks.

A screech will readily inhabit any box that fits somewhere between the Shackelford box design and a Wood Duck design with a 10 by 10 inch floor. Does any design attract screech owls better? Not in our experience. Availability appears to be a much larger factor.

Seven months after Pileated Woodpeckers nested in this tree cavity, an Eastern Screech Owl used it for a roost.

The OwlShack design (left) has spawned lots of imitators.

Screech-box openings come in various shapes and sizes.

A screech box is a wonderful DIY project and a great way to teach and/or learn some carpentry skills. The first sighting of an owl in your box is an incredible experience that is only enhanced if you built it yourself. One thing for novices to know is that the dimensions of store-bought lumber are not as they are labeled. For example, a common board to build a screech owl box would be labeled as 1 × 10 inches (thick × wide), but the actual dimensions are ¾ inch thick by 9¼ inches wide. Once you get the actual measurements of the wood you're using, then any design can be adapted with a little arithmetic. Also, this can be as much a creative natural art project as it is carpentry. Material such as cedar shingles or bark can be incorporated into covering a box to add a more natural look.

Entrance holes come in four general shapes: a round opening (3 to 4 inches in diameter), an oval (4 to 5 inches wide by 3 inches high), a slot (3 to 4 inches high along the width of the front of the box), and a larger squared entrance (5.5 inches across). As a general concept, the larger the opening, the deeper and narrower the box should be to keep the owls out of the reach of raccoons and other predators. A 3-inch-diameter circular opening is about the minimum entrance size.

There's no need for metal or slate hole protectors. Squirrels may enlarge the hole by gnawing the wood, but you should keep squirrels away from the entrance in the first place. A larger entrance won't affect screech use.

We've seen openings as small as 2.75 inches recommended, mainly to prevent more predators from accessing the box. Most screech owls may be able to use or nest in a box with an opening that small, and some research suggests that screeches tend to weigh slightly less in the southern part of their range. Smaller openings may still exclude a few owls, however, and a hole 3 inches in diameter or smaller will make it impossible to check on your owls with a larger cellphone and a selfie-stick.

We've seen a large female screech stretch a wing to fit into a 3-inch-diameter opening. Despite her struggle, she nested in the box. The extra-tight fit didn't stop her, but anything smaller may have. Why take the chance?

From a human perspective, the slot and larger square openings provide a better view of the owls and owlets if a pair chooses to nest in the box. It's amazing to see both the male and the female owl side by side or three to four owlets lined up all at one time in the entrance, and the slot and square opening are easier to build with just straight saw cuts—no hole saws are needed.

If you plan to install a nestcam above the opening, we recommend the slot design. When the male and female share the box early in their courtship, they sometimes like to sit in the opening at the same time around dusk. If only one owl can fit in the opening comfortably, the other may perch on the nestcam and literally try to kick out the other owl—and possibly move the camera in the process. If a nestcam is not a factor, oval and rectangular openings work equally well.

The roof is another key element. The roof tends to be the biggest weakness in nest-box construction. Untreated wood is certain to rot or crack with age, and the flatter the roof, the quicker the wood goes bad. You might want to give the top of roof several coats of clear polyurethane in the off-season. (The rest of the wood can be sealed, painted, or stained on the outside to increase durability, but never on the inside.)

Some store-bought nest boxes have peaked wood roofs. Over time, the wood can shrink, and the peak of the roof will leak like a sieve, prompting the owls to likely abandon the box. If your owl box has a peaked roof, cover the peak with a strip of vinyl or roofing material to protect it.

No owl box is completely weatherproof. Most store-bought owl boxes and reputable owl-box designs will incorporate ventilation and drainage. Some moisture will get in, so it is important to add a few small slits or ¼-inch holes below the roof for ventilation. If the floor of the owl box you've purchased lacks drainage, cut ½ inch off the corners of the floor or drill four to six ¼-inch holes. This enables the box and bedding material to dry out faster.

Chateau Ste. Michelle, a Washington State winery, sold a special collection in a small wooden case that could be converted into an owl box with some modifications.

Adding bedding material simulates the inside of a tree cavity and keeps the eggs from rolling around. Use dried bark mulch, playground mulch, crushed leaves, or pine needles. Wood shavings will do, but they can kick up enough dust to cloud the camera image for a day or so when a screech moves in. Avoid sawdust, which is way too dusty, and cedar chips, which emit fumes.

The bedding material should be 2 to 3 inches deep. A gallon Ziplock bag filled with bedding material is a good estimate for an 8 by 8 inch nest-box floor. Just empty it onto the floor of the box, and you're good to go. We've seen it as shallow as 1 inch, but it's best not to skimp. More than 3 inches of bedding provides diminishing returns, as the eggs may get covered in shavings, feathers, and debris, and the owlets may be within reach of larger predators.

Boxes should also provide access to the floor for replacing the bedding material. Various designs offer roof access, side access, front access, or a floating/removable floor. Each is a matter of personal preference, but keep in mind that some might be more challenging to use when standing on a tall ladder.

Barring nest disturbances, a once-a-year maintenance check of all things owl box related is a wise idea. If there was no nesting, replacing the bedding material is optional. If owls nested, replace the bedding after the owlets have fledged.

The nest box will be reminiscent of a chicken coop, with lots of feathers, decaying stuff, and a strong smell of ammonia. Let the box air out for a day or two before removing the old bedding. We recommend wearing rubber gloves, protective glasses, and a dust mask so the nest detritus won't get in your eyes. If you have a nestcam, now's the time to clean the lens with a microfiber cloth.

Some people add internal perches to owl boxes, which don't seem to attract or dissuade owls from nesting, but they come in handy when the female needs a break from larger jostling owlets. The perches also help owlets reach the entrance. A little ladder made of wire mesh (frequently used for Wood Duck boxes) can be added for that purpose. You can also use rough-cut lumber and make shallow saw marks or add scattered indentations by tapping a screwdriver with a hammer. Even with smooth wood, the owlets' wings should eventually be strong enough to get them to the entrance.

Similarly, an external perch at the entrance seems to neither attract owls nor dissuade them from nesting. Screech owls will use an external perch to sun themselves or make food transfers, but a perch can make it easier for squirrels and other predators to gain access, undermining any flashing that has been installed as a deterrent.

A Brief History of Nest Boxes

Bird boxes are big business these days. If you don't believe it, visit your nearest bird-watcher supply store and check out all the shapes and sizes. But these boxes have humble origins.

As far as we can tell, the first man-made nest box in Europe dates from the early nineteenth century. That's when an eccentric British nobleman named Charles Waterton built what is said to be the first birdhouse—and an owl box at that.

According to Adrian Thomas of the Royal Society for the Protection of Birds, the squire of Walton Hall "watched his birdlife through a telescope and decided that he wanted to encourage them to nest. So he set about making a stone box for Barn Owls to use. He didn't stop there. He built homes for Jackdaws, for Tawny Owls, and used fifty drainpipes to make nesting chambers for Sand Martins."

The Royal Society apparently never heard of North America's Choctaw and Chickasaw tribes. As a 2019 article in the *Austin Statesman* explained, members of these tribes "hung hollow gourds with an entrance cut into the side out on small trees and poles to attract Purple Martins, which in turn chased off crows, hawks, and vultures that might otherwise have helped themselves to the unguarded fields of corn or meat left out to cure. European colonists were quick to catch on and began putting out gourds of their own, as well as boxes, set against their houses for the birds, mainly to prevent hawks from carrying off their chickens." (As others have noted, the Purple Martins also came in handy when it came to killing mosquitoes.)

Building birdhouses hit the American mainstream in 1905, when the influential National Audubon Society's magazine *Bird-Lore* featured three articles on building them. One such article was by ornithologist Edward Howe Forbush, who wrote, "There is no better way to attract and protect several species of useful birds than to put up nesting-boxes. . . . While the exact size of the box is rather immaterial, the size of the entrance hole is most important. This should be just large enough to

HOLLOW LIMB BOX

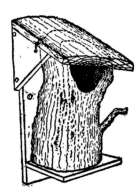

The Top of This House Is Held by Two Hooks.

Early screech-box designs based on the same idea—a hollow log with a roof and floor attached—appeared in 1905 and 1923.

admit the desired tenant, and small enough to keep out all larger birds." Forbush recommended 3.5-inch openings for two of those "useful birds," flickers and screech owls. But that was it.

The first article directly advocating nest boxes for screech owls came almost two decades later, in a 1923 article entitled "Bird Houses and Their Occupants" in the *Calgary Herald*. "On the whole [the screech owl] is much more a friend than an enemy of man," wrote noted Canadian birding writer P. A. Taverner. "It may sound inconsistent to allow a bird of prey to build in the garden, much more to advocate building a nest box for it, but it is an attractive fellow with many pretty ways."

Building a Squirrel-Resistant Screech Box

Scott built his first bird box in the sixth grade. He nailed together scraps of wood, climbed a ladder, and nailed the contraption to a tree. Squirrels probably used it until it fell apart that winter.

He got his first store-bought owl house as a grownup around 2000. Scott and his wife attached it to a pine tree at their first condo. Squirrels moved in by winter's end.

Fast forward to 2008. Scott had built several traditional screech boxes, and he decided to create a screech trail along the lines of a bluebird trail—a series of nest boxes that could be monitored regularly. His trail consisted of nine private-property sites, 0.5 to 2 miles apart, that might attract screeches.

A screech owl sits contentedly in the slot opening of one of Scott's owl boxes.

After researching screeches' nesting habits, he installed his owl boxes. The squirrels promptly moved in.

Do you see a pattern here?

Scott had to decide. He needed to spend more time monitoring the boxes to keep squirrels out (not a realistic option), use baffles or other squirrel deterrents on the boxes, or design an owl box that would stymie squirrels. He chose the last option. Note that we call this owl box "squirrel-resistant" not "squirrel-proof" because you can outwit squirrels only for so long.

Scott wanted a slick vinyl covering on the sides and on a steep roof that would keep squirrels from grabbing onto the box. He settled on a box that has a 16-inch-long roof (the wood is 15¾ inches long with vinyl flashing that extends ¼ inch beyond the edge) at a 45-degree pitch. A roof with less pitch would be less squirrel resistant, and a steeper roof would require more materials. Since the owl boxes are covered in vinyl on three sides, the box is accessed via the front panel, which can be attached with screws so that it pivots upward.

The resulting plans are pictured here. They're a bit complicated but not overwhelming.

A squirrel-resistant owl box before the flashing is added. Note the steep roof and pivot screws for front access.

You can choose an oval or circular entrance or a slot entrance. The round entrance holes look more old-school, but the slot entrance enables both owls (or a few owlets) to fit into the entrance at once. Another consideration is that making an oval and circular entrance requires the use of a hole saw, whereas a slot entrance only uses straight cuts.

The 14-inch-wide vinyl flashing is relatively easy to shape and attach to the box, but it can be chewed away if squirrels are really determined. Squirrels can't chew through aluminum or copper flashing, but these materials are much more difficult to work with and typically need to be nailed, not stapled. If you work with metal and have the tools and the know-how, go for it—a copper-flashed owl box would look regal. You can also install flashing on a standard screech box, but it will be less effective than on a steeper-pitched roof.

Squirrel-Resistant Owl Box (Slot Entrance)

WOOD: 1 × 10 inch board (actual ¾ inch × 9¼ inches), 8 feet long

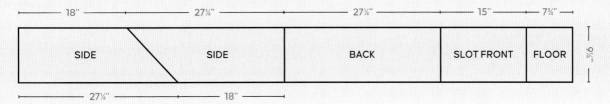

|← 18" →|← 27¼" →|← 27¼" →|← 15" →|← 7¾" →|

| SIDE | SIDE | BACK | SLOT FRONT | FLOOR |

9¼"

|← 27¼" →|← 18" →|

WOOD: 1 × 12 inch board (actual ¾ inch × 11¼ inches), 15¾ inches long

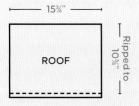

|← 15¾" →|

ROOF

Ripped to 10¾"

VINYL FLASHING: 14 inches wide × 6 feet long

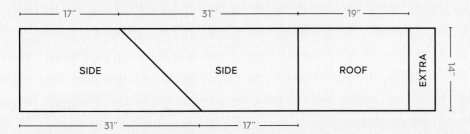

|← 17" →|← 31" →|← 19" →|

| SIDE | SIDE | ROOF | EXTRA |

14"

|← 31" →|← 17" →|

Figure 1. The blueprint for building a slot-style squirrel-resistant owl box. Note that the roof board needs to be ripped down to 10¾ inches to be flush with the sides.

Squirrel-Resistant Owl Box (Round Entrance)

WOOD: 1 × 10 inch board (actual ¾ inch × 9¼ inches), 10 feet long

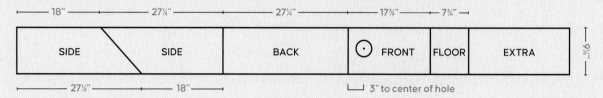

WOOD: 1 × 12 inch board (actual ¾ inch × 11¼ inches), 15¾ inches long

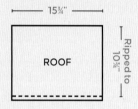

VINYL FLASHING: 14 inches wide × 6 feet long

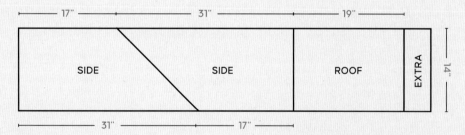

Figure 2. The blueprint for building a squirrel-resistant owl box with a round opening. Note that the roof board needs to be ripped down to 10¾ inches to be flush with the sides.

Materials

Pine boards are usually the least expensive, and the flashing will help them last. Cedar is a more expensive but a rot-resistant option.

· ·

For a box with a slot entrance

1 8-foot-long, 1 × 10 inch board (actual ¾ inch × 9¼ inches) for the sides and floor

1 15¾-inch-long, 1 × 12 inch board (actual ¾ inch × 11¼ inches) cut down to 10¾ inches wide for the roof

· ·

For a box with an oval or circular entrance

1 10-foot-long, 1 × 10 inch board (actual ¾ inch × 9¼ inches) for the sides and floor

1 15¾-inch-long, 1 × 12 inch board (actual ¾ inch × 11¼ inches) cut down to 10¾ inches wide for the roof

· ·

1 pressure-treated wood board, 1 × 4 inch and at least 32 inches long, for backing board to attach box to tree (or alternative brackets and hardware to attach to tree or post)

vinyl flashing material, 14 inches wide × 6 feet long

20 1½-inch-long deck screws

2 2½-inch-long deck screws (pivot screws)

1 2-inch-long deck screw (bottom front panel screw)

½-inch T50 type stainless steel staples to attach vinyl flashing to the box

a little bit of paint of a natural hue (optional)

Tools

pencil (to mark where to saw, cut, and drill)

tape measure

speed square (to mark straight and 45-degree angled cuts)

circular saw (one with a baseplate that can be angled to crosscut 45-degree angles is ideal)

hole saw with a 3-inch bit (for circular or oval entrance)

drill

screwdriver bit

rough sandpaper (for rounding the edges of the roof board)

strong scissors or tin snips (for cutting the flashing)

T50-type staple gun

hammer (to tap staples fully in)

paintbrush (optional)

These tools of the trade are used in building a squirrel-resistant owl box.

Construction Notes

- Follow good carpentry basics such as wearing safety goggles and predrilling holes for the screws that are smaller than the screw diameter.

- Store-bought lumber is typically smooth. Consider roughing up the inside of the front panel with very shallow saw cuts for owlets to climb to the entrance. A handsaw is probably best for this task. You can also add scattered indentations by tapping a screwdriver with a hammer.

Instructions

For either style of box:

1. Line up the upper back corner peak of the side panel with upper peak of the back panel. The back panel should be flush with the edge of the side panel all the way down the box.

2. Use three screws to attach one side to the back panel. Predrilling and inserting screws into a ¾-inch board takes some care; measuring ⅜ inch in from the edge of the side should center the screw into the edge of the back panel. Insert the first screw about 2 inches down from the top of the side panel, the second about 13 to 14 inches down, and the third about 24 inches down.

3. Repeat to attach the other side of the box.

4. Check to see how the front panel fits between the sides. Leave it in place unattached to keep the side panels aligned.

5. Use five screws to attach the roof flush with the back and sides of the box. Insert the first screw in the back top center of the roof, then insert two evenly spaced screws on both sides 3 inches down from the back edge and 3 inches up from front edge of the roof.

6. Cut ½ inch off each of the corners of the floor for circulation and drainage.

7. Recess the floor panel about ¼ inch from the bottom of the sides and the back of the box.

8. Use one or two screws each for the sides and back. Locate the center of the floor panel's edge, predrill holes for the screws, and then insert the screws.

• •

For a box with a slot entrance in the front panel:

1. Measure down 3 inches and 4 inches from the inner edge of the roof line on both sides.

2. Line up the top of the front panel with the 3-inch marks.

3. At the 4 inch marks, mark ⅜ inch in from the front edge of the side panels and predrill holes through the sides into the front panel for the 2½-inch pivot screws.

4. Attach the front panel with the pivot screws.

5. Close overhanging front panel. Near the inside center of the panel, make a pencil line where the panel meets the bottom of the floor. Open the panel and pre-drill a hole from inside to outside through the front panel ⅜ inches above the pencil line. This should help get the 2-inch front closing screw into the center of the floor panel edge to secure the front panel after bedding is added. (This is the screw that is almost fully unscrewed at the bottom of the open front panel in the photo on page 132.)

• •

For a box with an oval or circular entrance in the front panel

1A. For a 3-inch circular entrance, measure 3 inches down from the top and place a mark at the center of the front panel. Using a hole saw centered on the mark, cut a 3-inch entrance hole.

1B. For an oval entrance, measure 3 inches down from the top and place a mark at the center of the front panel. Move the hole saw ¾ inch to the left of center and partially cut a 3-inch hole, then do the same ¾ inch to the right of center. Once you have two partial cuts, complete both holes. Remove the two small triangular pieces of wood to complete the oval. Sand the hole well to make a smooth entrance.

2. Slide the front panel up until there is about ¼ inch between the top of the panel and the underside of the roof. Measure down 1 inch from the bottom of the roof edge and mark on side panel of box.

3. At that 1-inch mark, mark ⅜ inch in from the front edge of the side panels and predrill holes through the sides into the front panel for the 2 ½ inch pivot screws.

4. Close overhanging front panel. Near the inside center of the panel, make a pencil line where the panel meets the bottom of the floor. Open the panel and pre-drill a hole from inside to outside through the front panel ⅜ inches above the pencil line. This should help get the 2-inch front closing screw into the center of the floor panel edge to secure the front panel after bedding is added. (This is the screw that is almost fully unscrewed at the bottom of the open front panel in the photo on page 132.)

(1) Attach flashing to the sides first, wrapping from the front of the box to the back.

(2) Remove excess flashing that extends above the roofline.

(3) Ensure the roof flashing evenly overlaps the side flashing and that staples are positioned vertically.

(4) Wrap the roof flashing around the back of the box and secure it with staples.

Installing the Flashing

When attaching the vinyl flashing to the box, it's important to position the staples vertically. As the box ages, the staples will loosen and slide out to varying degrees. Horizontal staples may act as a grip or rung for a squirrel to gain access to the box.

1. Attach vinyl flashing to the sides of the box, moving from the front to the back and using your thumb and forefinger to crease/bend the vinyl around the edges. Allow about a ¼-inch overhang at the bottom to keep squirrels' claws away from wood and to prevent water from wicking under the flashing. Be careful not to impede the front panel from opening and closing. Use a hammer to tap in the staples fully, here and in the subsequent steps.

2. Cut excess flashing from the top back corners of the box.

3. Center the flashing on the front of the roof, allowing about a ¼-inch overhang at the front edge. Make sure the roof flashing evenly overlaps the side flashing all the way back. (This is probably the most difficult step.)

4. Fold the roof flashing at the back of the box, like wrapping a present. These staples can be horizontal, since they'll be positioned against the tree or post.

If you want to paint the flashed owl box, you won't need much paint. For a natural, camouflaged look, apply black, white, and warm-colored (brown to red/orange) paint to the vinyl with a brush. The natural wood front panel will turn gray as it ages and can be left unpainted. We also recommend painting the front edge of the roof and bottom of the box black to discourage carpenter bees from making egg tunnels.

Installing a Screech Owl Box

Once you purchase or build an owl box, you need to decide whether you're attaching it to a tree or a post or pole. We're not talking about the screws or wires—although you should use stainless steel or polymer-coated screws, which are less harmful to the tree—we're talking about box design.

In some designs, the back of the box itself is extended above and below and attached directly to the tree or post. Other boxes have metal T brackets or L brackets, or there can be eye bolts and a wire or small chain that wraps around the tree or post. We prefer to use a backing board.

This owl box is safe and secure and ready for nesting.

A backing board is a separate piece of pressure-treated 1 × 4 inch wood that is screwed to the box and used to attach it to the tree. We recommend adding a backing board or brackets to any nest-box design because storms and/or tree removal may generate debris that can damage or dislodge the box. If the nest box can break away from the tree without getting torn apart, it's less likely to get damaged. Replacing a board or bracket is easier than rebuilding a box.

In addition, the place where the owl box touches the tree bark is one of the first places to rot because it doesn't dry readily. Pressure-treated wood and/or brackets provide some breathing room (pardon the pun) for the box. Because of the chemicals, however, pressure-treated wood should *never* be used to build the box itself.

When it comes to where to place a nest box, you have two usual choices: on a tree or on a pole or post. Installing an owl box on the side of a building is straightforward, but some of the techniques we describe here may help make it easier and safer.

If you have a tree in a great location for an owl box, using that tree is probably the simpler, less expensive way to go. We have found two main drawbacks to tree installations: keeping out predators is more of a challenge because they can often access the box via nearby branches, and cleaning out the box after nesting season involves getting up on a high ladder (more on that later). A box on a pivoting pole or post is far easier to clean out.

A piece of 1 × 4 inch pressure-treated wood serves as the backing board on the nest box.

Safety First!

The idea of a screech owl nesting nearby prompts many of us to purchase a nest box. But once you get it home, the logistics of installing a large box 10 to 20 feet up a tree become daunting. Ladders, power tools, and heights, oh my!

Safety is the watchword. No owl box is worth breaking a leg over—or worse. According to the World Health Organization, falls from ladders cause more than 164,000 emergency room-treated injuries and 300 deaths in the United States each year.

- If you plan to set up a camera, it's best to install it in the owl box before mounting the box on the tree.

- Work with a buddy to hang the nest box, if you can.

- Place your ladder on a stable flat surface that is free of materials that could cause it to slip.

- Keep your center of gravity inside the ladder rails at all times.

- Never stand on the top two rungs of a ladder.

Installation on a tree

The most common way to install an owl box is to attach it directly to a tree, preferably 10 to 20 feet up. Look for a clean, vertical, fairly straight tree trunk or sturdy branch, with as few branches as possible below or around the box site. If needed, prune the trunk before installation. Branches near the box make it easier for squirrels and other climbing predators—like cats and raccoons—to gain access, rendering baffles or other deterrents useless.

If a clean, straight trunk is unavailable, you can try a few creative ways to secure the box in an upright position. If the trunk is slightly inclined, it's better to have the box face slightly downward. You may need a shim or other similar thin block of wood at the top or bottom of the box to hang it as close to vertical as possible. For larger angled tree trunks, you may need carpentry skills (think mini-treehouse creativity) to build brackets or supports for the nest box.

We have seen websites recommend installing an owl box on a large tree "as wide as the box" that would have a natural cavity large enough for screeches. The success of boxes placed on poles and posts belies this advice. Owls don't seem to care how thick the support for the cavity is, only that the cavity is large enough.

There are a couple of pragmatic considerations for installation on trees of smaller diameter. If the tree branch or trunk is narrower than your ladder at the top rung, be sure to secure the ladder to prevent it from sliding from side to side. Also, smaller trees are usually younger and faster growing, which means the tree can outgrow the box-attaching hardware quicker than on a larger, slower-growing tree.

You may need to prune branches below the intended installation site.

Although screech owl nesting has occurred as low as 6 feet, studies suggest that the owls prefer tree cavities at least 10 feet off the ground. Anything above 20 feet makes installation and annual maintenance more precarious. Generally, we recommend that the more activity the yard has, the higher up the box goes. Our sweet spot is from 14 to 18 feet.

Whether before or after installing the nest box, you'll need to secure some 18-inch-wide aluminum flashing around the tree to deter squirrels and climbing predators. Once you secure the flashing with clamps or long zip ties, you should paint it gray or brown to match the tree so it doesn't draw attention to the owl box.

The easiest way to get an owl box attached to a tree is to hire someone, typically a tree-trimming company. Just understand that this service will likely cost more than the owl box. (Alternatively, if you are an owl aficionado and are adept or already work with ladders, consider offering a reasonably priced service to install and maintain owl boxes through a local birding store.)

If you plan to install the owl box yourself, avoid using nails. Even relatively straight trunks have irregularities and varying bark thickness, meaning a 3-inch deck screw is usually good, but you may have to go a bit longer depending on the exact tree. Using a wire or chain (with eye bolts attached to the box) wrapped around the tree at the top and bottom of the box is also a popular technique, as this protects a little longer against tree growth pulling screws through the wood. One final creative way to account for tree growth is sandwiching washers and a spring onto each attaching screw. As the tree grows, it compresses the spring between the washer at the box and the washer at the head of the screw, allowing more time before the tree outgrows the attachment hardware.

No matter how the box is attached to the tree, eventually the tree will grow too much. Maintenance is key. Always check the attaching hardware when inspecting the box and bedding to see if the screws or wire need to be loosened. Check yearly and adjust when needed.

A simple way to hold the box in place for installation is putting a screw in the tree slightly above where you want the box to be attached and have a well-secured rope or metal loop at the top of the box. Slide the loop over the screw, and the box will hang like a painting while both hands are free to attach the box securely. Unfortunately, this method still requires carrying the box up the ladder.

A pulley is a great way to get the box into position without having to carry it up. The pulley acts like a second set of hands to hold the box in place while you attach it. A pulley that works with a ¼-inch rope, such as a clothesline, works well. Thread the line through the pulley and then temporarily screw the pulley to the tree above where the box will go or attach it by a rope or chain to a sturdy branch above. Once the line and pulley are in place, attach the line to the box with a hardware hook. Hoist the box into place, secure the line, and then attach the box to the tree.

A third way would be to recruit a friend or neighbor to help. Two people who are confident on ladders can set up an A-type frame, with each ladder on opposite sides of the tree where the box will be attached. One person holds the box in place while the other attaches it to the tree.

Finally, if the tree is suitable, instead of mounting the owl box directly to the tree, Paul Yorke of owlwatch.org recommends putting the box on the top end of a 1 × 10 foot board, using a heavy-duty wide hinge to attach the board to the tree, and then hoisting the box into place with a steel cable and winch. A latch secures the board in the vertical position. With this method, cleaning out the nest box and installing a camera is easier. (You can also use this method with a post.)

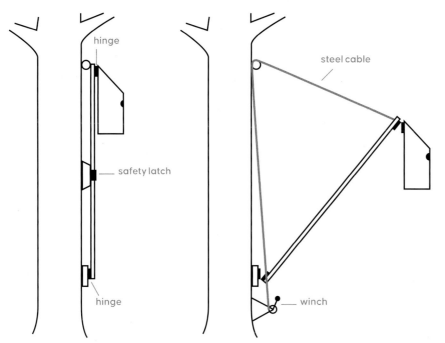

Figure 3. Plans for a hinged owl-box system.

Installation on a pole or post

We find that the following advice offered in a 1928 newspaper article still holds a bit of relevance: "Nest boxes should not be placed in trees, where they are accessible to the birds' enemies. They should be fastened to posts or poles. . . . Houses on poles are more acceptable to birds than others, probably because they are safer." While not quite as absolute as this statement, mounting an owl box to a post or pole has its benefits.

When installing a nest box on a pole or post, the same rules for location apply—including a tree canopy nearby. These installations cost more and require more effort, but they allow for flexibility in the placement and direction of the box, and they're a great way to keep out squirrels and other predators. They're also useful when your property lacks a suitable tree. Owls seem to find them perfectly acceptable.

Although some folks recommend using a Purple Martin nest-box telescoping pole for a screech box, we recommend securing it to a 2-inch steel pipe or a 4 × 4 inch (or even a 5 × 5 inch) wooden post. Screws are just fine for posts, but don't skimp on the length: use 2- to 3-inch screws. Metal poles will need 1- to 2-inch pipe straps that attach the pole tightly to the box with screws. Make sure screws don't protrude into the box, where they might harm the inhabitants.

Depending on how high off the ground you want the nest box, you can attach it directly to a pressure-treated 12-foot-long 4 × 4 inch wooden post, or you can sink a shorter 4 × 4 inch post into the ground and attach the nest-box post to that with two carriage bolts. First stain the post(s) with the appropriate shade so they blend into the surroundings.

The simplest way is to use the 12-foot-long 4 × 4 inch post. With a post-hole digger or shovel, dig a hole that's 3 to 4 feet deep. Attach the nest box at the top of the post, then attach a baffle halfway up the post to deter squirrels and other interlopers. Next, enlist a friend to help you raise the whole shebang into the hole, Iwo Jima–style. That's easier said than done, however, and it will be more difficult to clean out the owl box after nesting season.

Use a level to make sure the post is plumb, then backfill and compact the dirt around the post. It might be overkill, but you might consider using concrete to set the post in the ground for extra hold.

For a nest box that's 16 feet off the ground, use a post-hole digger or shovel to dig a hole that's 3 to 4 feet deep and insert a 10-foot-long 4 × 4 inch post. Attach the baffle so it sits at least 2.5 feet above the ground.

Next, with the help of a friend, use clamps to temporarily secure another 10-foot-long 4 × 4 inch post to the existing post, then drill two holes several inches apart through the posts for the carriage bolts. Insert one bolt, remove the clamps, and swivel the new post so it rests horizontal to the ground. Mount the owl box to the post so that it faces the side or upward—*not* face down. That way, when the nest box is lowered in the future, you can easily add wood chips and clean the box. Finally, swivel the post back into the vertical position and secure it in place with the other carriage bolt.

Above left: Use clamps to temporarily secure the nest-box post to the post in the ground, and drill two holes several inches apart to insert the carriage bolts.

Above: Use a ladder (or a friend) to help attach the second post.

Look Out Below

Once the box is up, check the ground underneath. This is where the owlets will land the night they fledge. Remove any rocks or other potential hazards. If the hazards are too big to move, strategically place outdoor chair cushions, towels, or something else soft under the box for the brief period when owlets begin to show themselves in the entrance—do so during the daytime when adults are less defensive—and then removed after the owlets fledge.

Relocating an Owl Box

Whether a box goes unused for numerous years, a tree or post is felled, or a squirrel usurps all defenses, sometimes it's necessary to relocate an owl box. As long as there is not an active nest in the box, fear not. We have moved boxes to another location 20 to 30 feet away with no problem.

If you put up the new owl box in a nearby location, we recommend removing the previous owl box as soon as possible to so the owl switches the new box and has time to get acclimated. If you see an owl in the opening of the old box, you can remove the box after the owl flies out. That way, an alternative hollow awaits when it returns.

If something awful happens (such as a tree falling with an active nest), it's best to work with a local rehabilitator to assess and address any damage and to reattach the nest box or replace it as close as possible to the original location.

Nestcams and High-Tech Help

Technology changes so quickly—for instance, solar-powered cameras are becoming more and more prevalent—that we're not providing any sort of buyers' guide in this chapter. Instead, we're offering the history and fundamentals behind the various types of cameras and other tools, using the equipment we've worked with as examples.

Nestcams can allow you to watch an often-amazing show few get to see.

Capturing Dame Nature's Secrets

Anyone having an earnest interest in both natural history and photography can find no more delightful and profitable way of spending leisure hours than by prying into the secrets of Dame Nature with an instrument capable of furnishing such complete and truthful information as the camera. It yields results that tell in no uncertain way of things and incidents that it would be well-nigh impossible to preserve in any other manner.

—Thomas S. Roberts, "The Camera as an Aid in the Study of Birds"

Nestcams

One of the first remotely operated nestcams was employed to photograph a Red-tailed Hawk nest in 1905, using a string tied to the camera's shutter. In an article for a Chicago newspaper, the anonymous author wrote, "The photographing of birds that build their nests in dark places necessitates the use of something which will make the place light enough for snapshots. Flashlight powders may be used, but the bright flash tends to frighten away the bird." His solution? "The reflection of the sun with a mirror is the easiest way to get sufficient light for quick work."

Three decades later, a remote-control nestcam that anyone could use was popularized by a syndicated newspaper article billed as a "simple method by which birds are made to take pictures of their own beauty." All a reader had to do was place a camera 2 feet from a nest and focus on where the bird would sit, then attach a string to the shutter, stretch it directly across the nest, and attach it to a stick on the other side. "When the bird 'comes home,' and sits upon the nest, her weight comes upon the stretched string, the shutter is actuated."

Researchers started using bulky VHS cameras in the 1990s to record avian activities. Mikael Kristersson's film *Kestrel's Eye* (1998) is an 86-minute documentary about a year in the life of a pair of falcons that nested in a church tower in a small Swedish town. The footage included many scenes filmed inside the nesting cavity.

In 2001 came *OwlCam*, a 100-minute video of Ward and June, a pair of Barred Owls nesting in the United States. The filmmaker used three cameras, including two in the nest, to document the lives of these large owls and their three offspring.

PHOTOGRAPH THEMSELVES.

Simple Method by Which Birds Are Made to Take Pictures of Their Own Beauty.

By the simple means here shown, birds are made to "photograph themselves."

The nest is discovered of a bird whose picture is to be obtained, and a foot or two away from and above the nest the camera to be used is fixed in a tree. The camera is then focussed upon the nest just as if a sitting bird there were being "taken."

Then comes an operation to be carried out with nicety. A string, at-

HOW THE CAMERA IS PLACED AND WORKED.

tached to the shutter of the camera, is carried down to and directly across the nest, which string is then fastened to a stem or branch on the opposite side. The picture makes all this clear.

And we see that when the bird "comes home," and sits upon the nest, her weight comes upon the stretched string, the shutter of the camera is actuated; and the gentleman in the picture shows us the subsequent result.

This 1905 article describes one of the first remote shutter release techniques for photographing birds.

"EXTRAORDINARY, AMAZING...
THIS BEAUTIFUL FILM ASTONISHES!"
-NEW YORK TIMES

KESTREL'S EYE

A Film By Mikael Kristersson

DVD

The Swedish documentary *Kestrel's Eye* was one of the first films recorded inside a nest.

As far as we know, this is the first documentary to use video recorded from inside an owl box.

Around that time, other nestcams were going online—notably a popular and long-running website created by Chris Johnson, a Texan with a nest box, a camera, nesting screech owls, and a detailed online journal. Another pioneer was the Cornell Lab of Ornithology. Tina Phillips started running the fledgling nestcam program in 1998. The inside of the nest boxes was very dark, so the staff retrofitted all the boxes with plexiglass tops so light could filter in.

"I was trained as a biologist, but I had never done this sort of thing," Phillips recalls. "It was 'learn by doing.' The first box we successfully had a camera in was a Tree Swallow box. In those days it was an image updated every 30 seconds. When I tell you these early systems were put together with duct tape, that's no exaggeration. We had no money in the budget, so we bought these cheap security cameras and ran wires 300 or 400 feet to an outlet."

"When we started, the Internet wasn't even a decade old," Phillips says. "When people were able to upload video more easily, that's when we were able to stream the video and pay for the extra bandwidth and storage. What we started with is nothing like the images we see now."

This is especially true of modern-day high-tech compact digital nestcams.

In Europe, for example, one of the most interesting projects was conducted by the Czech University of Life Sciences in Prague. In 2014 researchers developed a "smart" nest box with an interior that includes an infrared-light trigger, two video cameras, a microphone, temperature sensors, and a radio-frequency ID reader to log any tagged birds if they used the nest boxes. The front exterior panel is covered with aluminum flashing to keep out pine martens.

That's a far cry from 2006, when we installed our first camera. We ordered a camera and microphone from Supercircuits, a company that still sells surveillance equipment over the Internet. The $35 camera (Model No. PC6EX3, now obsolete), microphone, and power supply all required their own wiring and coaxial cables. Sound and video got plugged into a DVD recorder connected to a clunky television

The evolution of nestcams, from a 2006 soundless black-and-white camera (left) to today's color camera with built-in microphones.

set in the basement. The camera's infrared feature allowed night viewing, but the image was always in low-definition black and white.

In those early days, there was a tendency to sit and watch the nestcam on a TV monitor for hours on end during nesting season for fear of missing something. These days, thanks to livestreaming and various apps, you can watch what is going on from almost anywhere on a variety of other devices, including cellphones, tablets, and laptops. If you share the video feed on the Internet via YouTube, you can watch the action as it happens or rewind the video for up to 24 hours to see what you missed.

Timing of nestcam setup and privacy issues

One caveat from Robyn Bailey, project leader for the Cornell Lab of Ornithology's NestWatch program: "Owls are sensitive to disturbance, especially early in the nesting cycle. Set up a camera before nesting season, and then leave it alone. Don't put in a camera that's going to need the batteries changed every week. Maybe have an external power source or a solar panel because you want to disturb nesting owls as little as possible."

A small solar panel that contains a battery can power a security camera, while not disturbing the nesting owls.

One bit of advice regarding the equipment from HDOnTap, a streaming service for nestcams: "Our goal is to ensure that the equipment is weatherproof and waterproof. This isn't necessarily tough right away, but it can prove difficult over time. With extreme weather changes, equipment can fail at times. Working around audio issues can prove challenging as built-in audio for small IP cameras is not often very reliable. Troubleshooting issues can be difficult especially once owls are in the box. We can't do much to adjust the equipment without disturbance."

At some point, too, the lens on your camera will get dusty, water-spotted, or knocked askew. If you're within a month of nesting season or it has already begun, leave well enough alone. If you think you can tweak the camera without driving away your owls, you might go for it. Just ask yourself the question: What's more important, owls nesting in the box or getting the best images possible from the camera?

Don't worry that you're invading the owls' privacy. You aren't wiretapping the Kremlin or secretly video-recording fellow humans. Owls are intelligent in their own way, but there's no indication they have any notion of privacy and/or modesty. Our guess is that they would welcome a video camera with open wings if they knew it would help safeguard nestlings and keep out squirrels.

In a few instances, a nestcam has saved owlets' lives. Bill Powers, who started a network of eight screech owl cams, reports that at one nest in early May of 2017, "a Red-tailed Hawk was able to pull the female out of the box and killed her. We rescued the owlets and took them to this wildlife rehab center. The two owlets were released in the nesting area three months later."

Similarly, if anyone says that viewing the nestcam makes you a peeping tom, gently remind them that a voyeur involves unduly prying on other humans. All you're doing is viewing nature through a unique window. Nothing unseemly about that. If you're using an external camera, make note of the background to ensure you aren't inadvertently spying on the neighbors, and adjust the angle of the camera accordingly. An ounce of prevention will stave off misunderstandings between neighbors.

Choosing a nestcam

Avoid cheap cameras. Given the amount of time and money spent on the entire installation, this is the last place you would want to skimp.

These days nestcams come in all shapes, sizes, and prices, and they all offer sound, high-definition color images during the day, and infra-red images at night. Infra-red lights allow you to watch the owls in low light in black and white. A red-morph Eastern Screech Owl will often appear off-white, and the owls' eyes will appear off-white as well when they look at the camera, but the variety of owls that have nested successfully in boxes with infra-red illuminated cameras suggests that the owls are not adversely affected by them.

All nestcams also need a power source—even so-called wireless cameras—so keep that in mind when deciding which type of system works best for you. Live-streaming cams require the most power and usually require a power cord from the house to the camera. Solar-powered systems for these cameras are large and bulky—a 24-inch solar panel, plus a motorcycle battery for storage. Hopefully, the technology will progress quickly to make solar power a more pragmatic option.

The cameras are extremely useful for the same two reasons as their predecessors: to check the nest box for squirrel intrusions and to monitor the owlets' health. Once the camera is installed, you'll likely discover a third benefit—an around-the-clock look at what goes on inside an owl box.

Watching the nightly show during nesting season is must-see TV. Your friends and family will be fascinated, especially after the eggs hatch and the little fuzzballs squirm around. The tiny microphone seems to amplify all the owlets' screeches as they clamor for food. You can watch baby screeches go from eggs to hatchlings to fledglings. The process is a wonder of nature, filled with memorable moments and admiration for the parents, who work tirelessly to raise their offspring.

An added benefit of a nestcam is that as breeding season takes shape, your owls are more likely to bring in prey items, and you might become an expert at identifying the prey. When you can't, you should be able to ask a naturalist friend for help. *Peromyscus* mouse and *Blarina* shrew, anyone? (If you're wondering, owls prey on shrews—not at all bothered by the shrews' toxin that deters mammalian predators.)

The Bosch Dinion IP 5000i is recommended by Bill Powers and the streaming service HDOnTap.com, a free Internet service that can stream the video live over the Internet for you. Although the system requires adding a weatherproof compartment to house the camera and microphone, the image quality is excellent and the view is from the side of the box instead of the ceiling. The video streams directly to a television.

A more compact camera from Green-Backyard.com can be mounted directly on the ceiling of the owl box and wired directly to your Internet router via an Ethernet wire that provides power and connects the camera and its built-in microphone. Based on recommendations, we chose a Green-Backyard camera and, with help

This image was taken with a Green-Backyard camera.

2022-02-12 17:50:23

LocalHost

of an electrical engineer friend, installed it in early 2022. Two screeches occupied the box that winter, sometimes together, but did not nest there. Instead, a male Northern Flicker roosted there for a couple of weeks.

Others have recommended Birdhouse Spy Cam's Hawk Eye HD Nature Cam, which also streams the video directly to your television. When in doubt, shop around and ask around.

Watch a Screech Owl Almost Anytime

Thanks to the internet, you don't need your own camera to enjoy watching screeches inside their nest boxes. Although your chances are much greater during courtship and nesting season (roughly the first half of the year), one owl we follow roosts in the box from October to early June. Another roosts sporadically in the fall and winter, appearing in the opening near dawn or dusk.

Two excellent websites to watch screech cams and other wildlife cams are hdontap.com and PixCams on YouTube.

Security Cameras

Video cameras used for security (such as Nest and Blink cameras) can be installed inside or outside an owl box. They are not made for 24/7 livestreamed videos, but the videos can be viewed on demand with a cellphone from virtually anywhere that has a cell signal or Wi-Fi. Some security cameras also have settings for motion detection and recording—a handy feature. Since they are not designed to monitor birds, however, the cameras may click when switching between daylight and infrared mode or during photo captures, and they may have a small LED indicator light when in operation. It's best to test and disable any features that could be disruptive to nesting owls before installing a security camera inside an owl box.

Some camera systems have two-way audio, meaning not only can you hear what's going on via a microphone in the camera, but you can also talk through speakers in the camera. This feature is currently more common in security cameras than in those designed for livestreaming. You'll probably want to disable this

A security camera can be mounted inside an owl box.

feature if screech owls are present. You don't want to accidentally broadcast human voices or frightening sounds into an occupied owl box.

However, there are early indications that selective use of this feature—such as imitating a screech owl when a starling is in the box—might deter intruders. Further research is needed to determine what, if any, sounds might evict a squirrel. Perhaps a Red-tailed Hawk scream or a dog whistle?

Installing a security camera outside the nest box allows you to record the owls' nocturnal activities and to monitor squirrels and other wildlife near the box. The video output typically is not high resolution. The camera would also need to be close enough to your home to pick up your Wi-Fi signal. If not, you may need a Wi-Fi extender. Since security cameras are not in constant operation, they can be powered by a smaller (6 inches × 8 inches) solar panel containing a small battery.

Trail Cameras

A less-expensive alternative is a motion-activated trail camera that takes both still images and videos. If you want to record images or video of activity from outside the opening, you can place the trail camera (also called a trap cam) beneath the owl box. Trail cameras that record in infra-red without the pulsing red lights are less obtrusive.

The good news is you'll be able to see if a squirrel has sneaked into the owl box while you were away. The bad news is you'll probably have to approach the camera to retrieve the memory card—less than ideal if an owl is in the box. Trail cameras can also be powered by a smaller solar panel, and the internal batteries usually last long enough to replace when the owls aren't using the box regularly.

Cellphone Cameras, Selfie Sticks, and Mirrors

Opposite: This sneak peek inside an Eastern Screech Owl nest box was achieved with the help of a cellphone and selfie stick.

Below: Thanks to apps, you can view your videos on a phone from anywhere worldwide with cellphone reception.

If you want to check for squirrel invaders or sneak a peek inside your nest box, you can use a cellphone and a long selfie stick. We've seen Bluetooth-enabled selfie sticks that extend to nearly 10 feet, so depending on your height, that should do the trick. We've also read that others have used LED-lighted telescoping mirrors and baby monitors to the same effect.

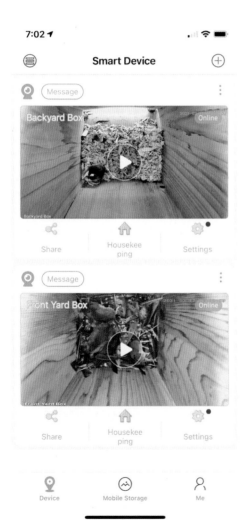

Apps

With more-advanced technology, you can watch the nestcam action via your cellphone, tablet, or laptop. You can record the nestcam video for later study, and you'll be able to livestream it on YouTube. Just be aware that if the owl nest is visible to the public and your YouTube livestream becomes known, you may have trouble keeping the curious and the photography-minded away. And that may cause the owls to abandon the nest.

Video Storage

In the old days, we recorded the video on a DVD burner. These days, everything is digital and takes up almost no space. But if you don't edit and delete rigorously, you'll eat up a lot of storage and have trouble finding the most memorable videos.

How to Install a Camera

Putting a livestreaming camera in your screech box will enrich your life immensely. However, if we said these cameras are easy to install for the average person, we'd be lying. The learning curve is steep, and since you're probably not installing a lot of cameras, it's a specialized skill that you'll seldom use. Installing these cameras is like doing complicated crossword puzzles.

When you're finally done, you know the answers, but you won't do that puzzle again anytime soon.

If you're ready to take the nestcam plunge and you know someone who has recently installed one, ask for their advice. If they like their camera, you might want to buy the same one. Even if they can't offer installation expertise, they can offer moral support. If you don't know a nestcam owner, ask your favorite handyman or electrician friend—or both—for advice. Or ask for help from a tech-savvy friend who will see installing a nestcam as a challenge, not a chore. Two heads are better than one when it comes to installing a camera for the first time.

If you hope to livestream nesting owls in early spring, we recommend installing the camera in autumn, if possible, and no later than the end of January. Earlier installation also gives house-hunting screeches more time to find the nest box, and it's easier to make adjustments. For a winter installation, begin early in the day because it can take several hours and you don't want to run out of light.

Another thing to keep in mind: if you already have an owl box and are installing another box with a camera in a nearby location, remove the nest box in the old location so the owls can't nest in it. Otherwise, all that work and wiring might be for naught—at least until the next nesting season.

The following is a description of how we wired a nest box in 2022:

Buying the camera

On the recommendation of friends, we ordered an Outdoor PoE Wired IP Bird Box Camera from Green-Backyard.com. PoE stands for "Power over Ethernet," a way to deliver DC power to devices over copper Ethernet cabling without the need for separate power supplies and outlets. An IP camera is an Internet Protocol camera, a video camera that transmits and receives data over a local network or the Internet.

The Outdoor PoE Wired IP Bird Box Camera offers several attractive features: it measures 1.8 inches tall by 1.8 inches wide by 1.7 inches deep, so it will fit easily in most nest boxes; you can livestream the video to any laptop, cellphone, or tablet; and it was relatively inexpensive (under $130 on sale, including shipping). With an additional app, you can stream the video in high-definition over YouTube.

The heart of this nestcam system is a microphone-equipped camera not much bigger than an engagement-ring box.

Solar-powered cameras with Wi-Fi are easier to install because they require only a small amount of wiring at the nest box. Note, however, that Wi-Fi will have some distance limitations from the network. A hard-wired camera will always provide a connection that will be more reliable and secure, and it will yield a better image.

Installing the camera

Expect glitches and snags, which seem to be a matter of course when installing any high-tech equipment for the first time. If you're ready to install the camera anytime soon, next week just might be "Take Your Local Electrician or Electrical Engineer to Lunch Week." Watching a YouTube instructional video can also be helpful.

Before you begin, make sure your Internet router is up-to-date and preferably located near where the owl box's Ethernet cable enters the house.

When the camera arrives, you'll open the box with great anticipation. It's more than just a box of wires, plugs, cable, and a camera the size of a 3-inch cube. It is a little box of possibilities that, if things go well, might offer some of the most amazing images you will ever see. Then you'll read the instructions and wonder why the manufacturer didn't include a bottle of aspirin with all the wires and gizmos.

A few things we learned the hard way. Download the app and test the camera before you install it in the nest box and before you install the box on a post or tree. The camera comes with a 4-foot-long test cable. Use it to connect the camera to the router while you're inside (especially in colder climates) and can get everything working. Similarly, installing the camera in the owl box is the easy part—so long as you do it inside your home or workshop. You don't want to attempt this on a ladder outside unless you have no other choice.

Using the swivel bracket, attach the camera as near to the peak of the nest box's ceiling as possible. Aim the camera at the floor of the box and check it on the video feed. To make sure the lens was in focus and aimed where we wanted it, we put a wind-up toy monkey about the same size as a screech owl in the box.

This wind-up toy monkey was the stand-in for a screech owl when we tested the camera's image quality and focus.

Be sure to screw the swivel bracket to the ceiling as securely as possible—screeches jostle cameras during nesting season. If your screech box has a steeply pitched roof and there's room for an owl to perch on the camera, add a block of wood just above the top of the camera or frame it on three sides. Otherwise the owls can change the carefully crafted point-of-view of the camera.

Drill a 1-inch hole in the back of the nest box near the camera. The hole will be big enough to accommodate the connector to the Ethernet cable, which powers the camera and transmits the sound and video. Once the cable is connected to the camera, run the cable through the opening, cover the rest of the opening with duct tape, and plug it with a sealant like Flex Paste. (The duct tape keeps the goop in place until it hardens.)

Decide where the Ethernet cable will enter your house (typically a basement window), then measure the distance from the top of the nest-box pole all the way to the window, including every twist and turn along the way. Measure the distance again—it's always a good idea to measure twice. Also measure the distance from the basement window to your Internet router via your basement ceiling.

Err liberally on the side of caution. If your measurements come up short, you'll have to buy a longer cable. Splicing Ethernet cable is a real challenge. If your measuring tape isn't long enough, use twine to mark the entire route, and then measure the length of twine.

Our installation needed approximately 150 feet of cable to go from the owl box to the router. The easiest way was to buy a 100-foot spool of cable with connectors at each end to connect the camera to the surge protector, and a 50-foot spool of cable to connect the surge protector to the Internet router.

Running the camera cable

Run the Ethernet cable down the pole or tree trunk, securing it with staples. (Some folks advise spraying the exposed wire with pepper spray to discourage squirrels.) If possible, dig a shallow trench from the pole or tree to where the cable will enter the house, and then bury the cable so squirrels are less likely to chew through it and folks won't trip over it. Long term, you might want to run the cable through a 1½-inch-diameter PVC pipe for better animal-proofing and weatherproofing.

Don't forget to install a surge protector with a ground wire.

The basement window is a good place to install a surge protector to safeguard the camera in case of a lightning strike.

Inside the house, you can thread the cable above the basement ceiling to the router. This may involve lifting some ceiling tiles. Remember, slow and steady wins the race.

When you're done, we recommend livestreaming the video feed through You-Tube, which not only enables other folks to tune in, but it also allows everyone to rewind the feed for up to 12 hours to see any action they might have missed—like a screech dropping off a prey item during the night. Our friends recommend a free download called OBS Studio, which takes the signal from the camera and converts it to livestream on YouTube. You'll also need a Google account. If you are new to livestreaming, you should get a tech-savvy friend to help—it's a bit like learning a foreign language that you'll seldom use.

When a Screech Owl Arrives

How soon an owl will move into your nest box is a bit of a crap-shoot. It could be days—or years—even if you do everything right. The more you do your homework on site selection, the better your chances will be.

When you see an owl in the opening for the first time, savor the moment—but keep your distance and pay as little attention to the owl as possible. You can still do yard work, but don't stare at your visitor. If you want to photograph it (and who can blame you?), use a telephoto lens and avoid using a flash.

The first few times an owl shows up are likely a trial period. How suitable this hollow is may depend on the owl's other options. Thus, although you will be most excited about that first sighting, discretion and distance may help the owl gain confidence in your property as a safe space.

In the north, screech owls sometimes settle
into a nest box in winter.

Each owl has its own personality. Some will seem very skittish and drop down into the box at the slightest sight of movement. Others will comfortably watch the entire lawn being mowed. A narrow screech with erect ear-like tufts in the box entrance is in a defensive alarmed posture, even if its eyes seem closed (especially during the day). It's time to move away and train your gaze elsewhere. At times a screech will wait until eye contact is broken to tuck back into the box.

The likeliest times for the owl to perch in the opening are dawn, dusk, and when the sun hits the box on cold days. It's perfectly normal for a screech to roost in various places, so don't worry if it disappears for days on end. Similarly, it may choose another location to nest. Don't blame yourself if the owl doesn't come back. That's life.

A Most-Familiar Owl

A weird, melancholy, whistled tremolo from under our very windows startles us, as the uncanny voices of all owls do, however familiar we may be with the little screecher. Are any superstitions more absurd than those associated with these harmless birds? Because it makes its home so near ours, often in some crevice of them, in fact, in the hollow of a tree in the orchard, or around the barn lofts, this is probably the most familiar owl to the majority of Canadians and Americans.

When hunting, the owl moves like a shadow, so silently does it pass in the darkness.

—Neltje Blanchan, *Birds That Hunt and Are Hunted*

Before the Hatch

Courtship typically involves male displays: bowing, raising his wings, clicking his bill, and bringing food to the female to show that he's the avian equivalent of a breadwinner. Consider this 1892 screech owl courtship description from the classic *Life Histories of Familiar North American Birds* reported by Dr. Lynd Jones:

> I saw this species mating once. The female was perched in a dark leafy tree apparently oblivious of the presence of her mate, who made frantic efforts through a series of bowings, wing-raisings, and snappings to attract her attention. These antics were continued for some time, varied by hops from branch to branch near her, accompanied by that forlorn, almost despairing wink peculiar to this bird. Once or twice I thought I detected sounds of inward groanings, as he, beside himself with his unsuccessful approaches, sat in utter dejection. At last his mistress lowered her haughty head, looked at and approached him. I did not stay to see the sequel.

Screech owls are said to mate for life. If the female dies in the territory, the male will find a new mate—and vice versa—and remain on that territory for life. Owl researcher Jim Belthoff, long associated with the Raptor Research Center at Boise State University, reports an instance where one female stayed on a territory for 13 years and bred with several males.

In the run-up to nesting season, a male screech may rotate among prospective nest sites to roost to protect his turf. Sometimes the male will roost mostly in one location, but you needn't worry if you fail to see the male in the nest box daily. Some screeches are so stealthy that owl box owners don't realize their nest box is occupied until they see owlets in the entrance just before fledging time—one of the many reasons to install a camera in the nest box.

The owls' next step is to choose the nesting site. Lisa Potash of northern New Jersey has two owl boxes with cameras on her property. Lisa has watched the male screech owl show the female possible nest sites before she chooses one.

"Several times during courtship, I observed a male enter one owl box, then the female would enter," says Lisa. "Most often the male would then fly along the edge of the yard to the second owl box to roost for the day. Sometimes they would go in and out of a box multiple times—predawn and sunrise—and switch boxes multiple times or end up together.

Screech owls start staking out territory for the nesting season as early as autumn (painting by Louis Agassiz Fuertes).

2022-02-12 16-26-23

LocalHost

Above: Part of the courtship routine can involve sharing an owl box and preening each other, known as allopreening.

Right: An early sign of breeding season is prey items in the screech box.

"Surprisingly, I observed how a male would not 'allow' the female entrance into the box where he was perched at the opening at dawn. After making bill-to-bill contact, the female would then fly to the second box for the day."

If you installed a nestcam, you may see a screech bringing in dinner, such as a bird, in late winter and tearing into the carcass with the vigor of an in-law on the day after Thanksgiving. As soon as this happens, that pristine carpet of wood chips will resemble a chicken factory floor at the end of the night shift. Although this mess looks gruesome at first, we're guessing that you will grow so accustomed to it that you may soon try to identify the bird meals by the feathers left behind.

Your owls may use some of the feathers as weather-stripping, cramming them into any gaps between the clean-out door and the rest of the box. (If you don't have a camera and notice feathers sticking through the gaps, that's a sign you have a screech for a tenant.) To watch the feathers accumulate is to understand why other birds hate screech owls so much.

You may see the owls in the entrance consistently around dusk or dawn, and then the box might go curiously quiet. That may mean the owls chose to nest elsewhere, or it could also mean the female has started incubating her eggs—during which she spends hardly any time in the entrance. Unless the male is needed to defend the box from an imminent threat, he usually will be roosting somewhere nearby but not in the nest box itself.

When you see a screech roosting in the nest-box opening in late winter, don't feel compelled to avoid the area. Without making direct eye contact, walk at an angle maybe 30 feet from the owl box and softly say "hi" in hopes of acclimating the owl to your presence so you won't be perceived as a threat.

Tony Angell, the author of two books on owls, also became a familiar sight to the owls nesting in his yard. As he noted in an email, "That I could be present around the nest box and not disturb the pair of birds demonstrated to me the owls' capacity to identify individuals. I had established myself as a non-threatening human. Not everyone who visited the birds enjoyed such confidence, and the owls would keep their distance or even scold intruders. My daughters seemed to have an immediate rapport with the owls—almost as if the adult owls could distinguish that little children were less threatening than a grown-up human."

Owl Eggs as Time Capsules

Besides a nest box, the only place you're likely to find legal screech owl eggs is in a natural history museum. The Western Foundation of Vertebrate Zoology in Camarillo, California, houses more than a million eggs from more than 4000 bird species. It's the largest collection of eggs—including some from the nineteenth century—in the world.

A 2020 report published in *The Auk*, a leading ornithology magazine, described bird eggs as "instantaneous, well-preserved snapshots of bird breeding that contain information about past environments and hold unique biological information not available [elsewhere]."

For example, because of the effects of modern chemical pollutants on the composition of eggshells, older eggshells help document the

These Western Screech Owl eggs were collected in California in 1896.

long-term changes in the environment. By comparing the calcium levels in peregrine falcon eggs collected before the advent of DDT to eggs collected during the DDT era, scientists proved that the pesticide was making the shells thinner and more brittle. The resulting public outcry led to the use of DDT being banned in the United States, allowing for a dramatic increase in raptor populations across the country.

Incubation

Although screeches live in an area year-round, the one period they will use an owl box continuously is from around when the first egg is laid until the last fledgling leaves—covering roughly 2 months. (If you're lucky, one might occupy a nest box for 8 or 9 months of the year.)

Screech owl breeding season can begin as early as January in Florida and other places with warm year-round climes, and a nestcam documented an instance in Texas where a screech laid eggs in mid-June—very late, even in the northern part of the range. On average, though, mid-March through most of April is prime egg-laying time for Eastern and Western Screech Owls throughout much of their range.

Owl eggs tend to be rounder than the eggs of most other bird groups. Screech owl eggs are semi-glossy, a bit granulated, and roughly 1⅛ to 1¼ inches in diameter. Like the eggs of most cavity-nesting birds, the eggshells are white—or "whiter than even moonlight," as the author Mabel Osgood Wright once described them.

The theory behind that lack of pigmentation is that because the eggs are out of view of predators, they don't need camouflage. At first glance, the eggs resemble ping-pong balls, just smaller and perhaps a tad out of round.

Once the first egg arrives, expect an additional egg about every 2 days, give or take, with three to five eggs being the typical range. On one occasion, 6 days had elapsed before the fourth (and final) egg was laid. For those with nestcams, the biggest moment of the day is the big reveal, when the female leaves for a potty break and you can count the number of eggs.

Although the eggs in a clutch look identical, science suggests that they are not created equal. D. Caldwell Hahn, now a professor emeritus at the

This ping-pong ball (left) is a bit larger and rounder than the replica screech owl egg next to it.

Front Yard Box

2022-04-12 17:19:36

A major nesting-season milestone is the arrival of the first egg. Note the ample feather insulation around it.

Eastern Ecological Science Center in Maryland, studied the yolk hormones within clutches of eggs of Eastern Screech Owls. She found that the levels of yolk androgens, akin to steroids, "were lower in first-laid eggs, which may reduce dominance of the first chick." Further research has shown this occurs with other bird species as well.

Some owl books state that incubation usually begins with the arrival of the first egg, but we've found that incubation begins once there are two or more eggs. First-time nestcam owners often get concerned when the female screech leaves the nest box for extended periods at night after the first egg or two arrives. There is seldom a reason to worry—even if the weather is extremely cold—because incubation has not begun.

Occasionally, if an egg is laid early in the season or a late Arctic front moves in, an egg will freeze before incubation has started, making it inviable. This and the occasional unfertilized egg can be a cause for concern, but they are just part of the natural processes. The remaining eggs usually proceed as anticipated.

During the breeding season, the female develops a brood patch, a small featherless area on her chest that's well-supplied with blood vessels. The brood patch makes it easier to transfer heat to the eggs during incubation.

Once the female starts incubating, she sits on the eggs most of the time. Incubation takes roughly 4 weeks, with the male providing food at night. That means there is very little activity to be seen during the day from outside the box, aside from Blue Jays and smaller birds flying to the opening and occasionally announcing their annoyance at the owl's presence.

With screech owl eggs, four usually means a full house.

Front Yard Box
2021-04-14 19:50:27

Inside the box, the female is mostly still and prone, moving occasionally to preen, feed on any cached prey, or reposition herself and/or the eggs. Two exceptions are when she flies up to the opening to catch a little sun or to get food from her mate, and when she flies out after dusk on a potty-and-pellet break, which can take 20 minutes. If you have a nestcam with a microphone, you will sometimes hear her call for food—seemingly impatiently—beginning around dusk.

Toward the end of the fourth week, nestcam owners should be on the lookout for a small hole in any of the eggs. This is the start of a process called pipping, in which the chick breaks through the egg and emerges as a nestling. Pipping takes up to a day or two. And then it's on to the next stage, developing in the nest instead of the egg.

Ever wonder what happens to the eggshells after the chicks hatch? The mother owl shreds and eats them to help regain the calcium she has lost. (Which came first, the screech owl or the egg, is a topic for further research.)

If you don't have a nestcam, discreetly watch for the male. He typically roosts nearer to the nest box when the eggs are about to hatch and has cached prey items nearby as well. Research suggests that the leading cause of nest failure is not predators, as one might suspect, but the failure of eggs to hatch. Robyn Bailey, who heads Cornell's NestWatch project, says that unhatched eggs may indicate that something befell one of the parents. The female would abandon the eggs if she had no help from her mate.

When an egg goes missing

Many years ago, we received a report from a New Jersey birder with a nestcam and a female owl sitting on eggs. One night, the egg count went from four to three. The birder was trying to figure out what had happened. The male was also spending a lot of time in the box, and the birder wondered if one of the adult owls had inadvertently broken an egg.

For answers, we consulted two owl experts with plenty of knowledge and first-hand screech experience. Their verdict was theft by predator.

The first expert was brief: "I'm going to guess that what is happening is he has a predation problem . . . one egg missing and the male is in the box to help defend the nest."

The other expert was more expansive: "I rather doubt that by jumping into the nest box, the owls would accidentally damage the eggs. Look at their investment in all of this. Rather, I would look to red squirrels, gray squirrels, weasels, or snakes as possible predators. The calcium (as well as the protein) in the eggs is an important factor for these animals, who are also trying to gear up for the breeding season."

He also warned about future raids, warning that "the predators remember these boxes. In Europe, it was found that some of the predators (martins) remembered, from year to year, where their 'food boxes' were, and made it a normal part of their visit schedule."

And that's exactly why squirrel-resistant screech boxes and predator guards are wise investments.

Otus, by Robert McClung, introduced young readers to screech owls in the wild.

Otus

This hard-to-find 1959 children's book, with illustrations by Lloyd Sandford, depicts a nesting season of a screech owl with incredible accuracy and detail. The owls have no human characteristics. They don't talk like humans or display human emotions. They do, however, live precarious lives as both predator and prey. In two scenes, a cat and a fox kill an owlet, but the events are presented without sentimentality. They are just another aspect of nature and a fact of life—something that anyone who has watched a nestcam has likely witnessed.

Naming Owlets

Many people believe it's bad form to name wild animals as though they were pets. Others say it's a free country and you can name them as you choose. Who's right?

The argument against names: Owls are wild animals. Giving them names trivializes them and makes them seem more human or pet-like and less dangerous. That distinction is especially important around fledging time, when the parents will become aggressive in defending their young.

As syndicated columnist Froma Harrop has argued, "We grow up with teddy bears and stuffed lion toys. Ideally, children—and adults—will learn to distinguish between make-believe and biology. Holding back on giving wild animals names might be a good start."

The argument for names: If you have a pair of screeches living in your yard, referring to them as the "male screech" and the "female screech" for weeks on end is impractical. These owls almost seem like neighbors. Do you refer to someone next door as the "male human" or the "female human"?

Author Robert McClung sidestepped the problem in his wonderful 1959 children's book *Otus* when he gave his title character the Latin name for a horned owl. These days, one might compromise and call Eastern Screech Owls "Ace," short for the Latin *Asio*, and Western Screech Owls "Kenny," short for the Latin *kennicottii*. Or just plain "Owl."

We have a final word of advice. Just as you should never count your chickens before they hatch, don't name your owlets before they fledge. Hope for the best, but accept that some owlets will never make it out of the nest. Naming them will only make you feel worse if they don't survive.

After the Hatch

One of the most exhilarating moments in nature is seeing just-hatched owlets for the first time. They are delicate little fluffballs. The owlets make equally tiny sounds, halfway between peeps and gripes, as they ask to be fed. Watching owlets bring the words of Aristotle to mind, "He who sees things grow from the beginning will have the best view of them."

The male typically brings in the prey, and the more the merrier. The female's job is to feed the young. At times, she'll feed them smaller insects directly. Other times, when rodents are in plentiful supply, she will cache them, tear them into bite-size pieces, and feed them to the owlets as needed. If you have a nestcam, you might see the male feed the owlets bite-size bits of a rodent directly, but that's an exception to the rule.

The owlets soon sprout little wings and grow in leaps and bounds. Some say the owlets in their white feathers resemble lambs at this point, while others say they look like dandelions about to disperse their seeds. The female still flies out after dusk for her potty-and-pellet breaks, and the owlets often huddle together for warmth like some sort of rugby scrum.

Three baby screeches huddle next to the last egg to hatch.

During warmer days when the owlets no longer need their mother for thermoregulation, they can be quite rambunctious. The female often will escape the fray in the box to roost in the foliage, but quite close—usually within 30 feet of the entrance of the nest.

The owlets' feathers grow slightly darker by the day, so they will blend in better with their surroundings after they leave the confines of the nest box. They look less and less like babies and more and more like gawky kids.

The closer it gets to fledging time, the bigger the owlets get—to the point where there's scarcely enough room for the female inside the box. One or two of them may be larger than the others. Research suggests that while screech nestlings aren't aggressive toward one another, they do establish a dominance hierarchy about who might get fed more. With

Three owlets stare at a bug while their mother perches above the fray.

other owl species, large discrepancies in owlet sizes have led to problems during times of food scarcity.

If you watch on a nestcam, you can see a nightly drama unfold. The owlets wait for a parent to return with food, with the bigger owlets acting like NBA power forwards blocking out smaller players from getting the rebound.

It's feeding time.

The parents fly in with dead moths for the owlets, and the bigger owlets always seem to snare them at first. But the adults make sure the smaller ones get fed, too, flying in later with bigger prey, tearing it into pieces and putting it into the mouths of all four nestlings.

One of the most fascinating times is the week before fledging. Not only do you marvel at how big the owlets have gotten, but you also wonder how they all still fit inside that nest box. When they sleep, they resemble sardines in a can sometimes. Parents and owlets become active when daylight fades. The owlets stretch their wings and screech more loudly, and you might soon glimpse one or two in the owl-box opening.

On mild spring evenings around fledging time, parent owls spend a lot of time catching and feeding insects to the young. It takes a lot of insects to feed those hungry mouths, so parents fly to and fro quite frequently and can average a food run every minute or so.

Four Eastern Screech owlets peer from the nest-box opening.

This is when the activity in and around the box can be so captivating to observe. You might try setting up a few lawn chairs comfortably away from the box and off to the side, and then sit still as sunset approaches. With a little luck and quiet patience, you might witness the screeches doing their incredible owl thing. Owlets are also quite vocal begging for food at this time, and parents can respond in sometimes barely audible tones.

Predicting when the owlets will take that leap of faith one night and jump from the opening is mostly guesswork. Sometimes, the owlets leave one at a time, night after night, and other times they all take the plunge the same night. If an owlet overstays its welcome in the nest box, the parents will stop feeding it and force it to take that big leap.

This owlet is hoping for a meal delivery.

Once the owlets fledge, they don't return to the nest box. If you have a nestcam, one of the biggest shocks of watching a screech family go from incubation to fledging is how, seemingly in the blink of an eye, the owl box shifts from Bedlam Manor to utter silence.

The owl box will likely remain empty for several months, so be prepared to deal with real-life empty-nest syndrome. Keep reminding yourself that someday soon you just might see a familiar screech owl looking out of the opening of that shoebox-sized rectangle at the edge of your yard, and that roller-coaster ride will begin anew.

A Brood of Screech Owls

At one point in the grayest, most shaggy part of the woods, I come suddenly upon a brood of screech owls, full grown, sitting together upon a dry, moss-draped limb, but a few feet from the ground. I pause within four or five yards of them and am looking about me, when my eye lights upon these gray, motionless figures. They sit perfectly upright, some with their backs and some with their breasts toward me, but every head turned squarely in my direction. Their eyes are closed to a mere black line; through this crack they are watching me, evidently thinking themselves unobserved. The spectacle is weird and grotesque and suggests something impish and uncanny. It is a new effect, the night side of the woods by daylight.

—John Burroughs, *Wake Robin*

The Owlets' Next Adventure

You may wonder where young screeches go after they fledge. "When they first leave the nest box, they tend to select roost spots that aren't that camouflaged," says Jim Belthoff, a professor in the Department of Biological Sciences at Boise State University who has studied Eastern and Western Screech Owls for several decades. "They're not very mobile, so they just end up where they end up, and it's easier for the parents to feed them out on an open limb. Over the next few days, they become more mobile and start selecting those tangly-vine kinds of roost sites."

Belthoff notes that an accumulation of whitewash under an open limb is a telltale sign that a young screech owl has been there. There's also an accumulation of whitewash under limbs where screech owls have roosted as a family. "I look down to find them as much as I look up," he says.

According to Belthoff, the offspring of both Eastern and Western Screech Owls spend about 8 weeks in their natal territory and seldom use the same roost

Opposite and below: After nestlings fledge, they often hide in nearby trees.

site on successive days, perhaps to thwart predators. Distances between the roost sites also increase toward the end of the post-fledging period, likely reflecting the young owls' increased flying skills and independence.

During that time, the young tend to roost away from the adults more. In late July or early August, they disperse and head out to find territories of their own where they can survive the winter, settle, and try to attract a mate the following spring.

The adults remain on their territory. The weeks after the young disperse is a peak time to hear screech owls vocalizing. If you keep the windows open at night, you might hear their eerie calls.

When Screech Owls Attack

After following these owlets so closely for so long, you may be tempted to try to find out where they are lurking. You're better off leaving them alone. You'll only stress them out and stress out their poor parents, who must safeguard them and feed them. This is also the time that humans need to be most cautious in the vicinity of the nest.

Screech owls may look adorable, but don't let appearances fool you. Although screeches tend to live near humans, they will defend their young with gusto when people get too close to their nest or fail to heed their warnings (typically by loudly clacking their bills). No wonder Mary Oliver, in her poem "Little Owl Who Lives in the Orchard," described the screech owl both as "a memo from the offices of fear" and a message "from that mysterious conglomerate: Oblivion and Co."

As a Canadian newspaper columnist advised long ago: "For the most part, screech owls are harmless,

Birds in Our Lives, a 1966 book published by the US Department of the Interior, featured a rare photo of a person getting strafed by a screech owl after removing an owlet from the nest.

SCREECH OWLS ATTACK PEOPLE IN COLUMBIA

Harrisburg, Pa., Today.— (AP)— A breed of screech owls that attacks people with beak and claw as well as with wings seems to have developed at Columbia, Lancaster Co.

Deputy Game Warden Jerome Culp has reported to the State Game Commission that he had shot three of this protected species upon complaint of numerous residents. The owls had made their homes in trees near Mount Bethel cemetery and made several unprovoked attacks upon citizens who ventured near.

Joseph Gamble, sexton at the cemetery, said he received cuts on his forehead on two occasions and three other persons had their skin broken by the owls.

Dr. Thomas E. Wincroff, head of the bureau of research and information of the commission said he knew of screech owls striking persons with their wings but had never heard of them using their beaks or claws. Several Columbia citizens confirmed Culp's report.

It's th' Hoot!

Screech Owls Attack on Night Strollers

ST. CHARLES, Mo., June 28 (AP) —Police blame the heat for the town's latest menace—screech owls.

The ordinarily harmless birds have been on a rampage during the extreme warm weather, swooping down and scratching unwary evening strollers.

Police shock troops have bagged 13 owls; so now the citizenry is breathing easier.

Telephone Repairman Injured in Battle With Owl During Sun's Eclipse

One fatality and one casualty in South Pasadena were directly traceable to the eclipse of the sun Monday forenoon. The dead was a screech owl, lured from its day-time haunt in the large trees near Oak and Marengo avenue, by the semi-darkness. The injured was Henry N. Ireland, repair man for the Southern California Telephone Company. The tragedy was the sequence of an aerial battle which the telephone man waged with the night bird.

Mr. Ireland was working on a telephone pole at 1601 Marengo avenue. The owl used to having its own way during the hours it was abroad, resented the presence of an intruder. It attacked the repairman with a rushing flight sinking its beak and talons into his face. On its next rush, he struck it with his hand, killing it. Following the battle, the telephone man climbed down from the pole, took the owl as evidence of his strange encounter and went into headquarters for first aid treatment.

These are just a sample of owl attacks reported in the press.

except in late June or early July, when Ma Screech is teaching her chicks to fly. Then beware and steer clear of the nest." Sound advice.

In researching this book, we reviewed more than a century of newspaper accounts of screech owls attacking humans. The good news is that the violent attacks were so few and far between that they were newsworthy. Virtually all the incidents involved humans venturing too close to a screech owl just after the young had fledged. Often, the attack was attributed to other reasons, including a solar eclipse—never mind that the attackee was a persistent telephone repairman working near a stand of trees where the owl likely had a nest.

In many cases, sadly, the offending owls were killed. For example, in 1933 in East Alameda, California, a policeman shot a screech owl after many neighbors reported the owl was making "a terrible noise." Another screech was shot and tested for rabies, even though birds can't get rabies.

The most notorious owl attack in history is likely the one involving British nature photographer Eric Hosking. It occurred back in 1937, when camera technology was still young and photographers didn't seem to know any better. In his autobiography, the aptly titled *An Eye for a Bird*, Hosking wrote that "no nesting season is complete without photographing owls." Thus, when he heard about a pair of nesting Tawny Owls in Wales, he devised an elaborate plan to photograph them.

Using 25-foot-long poles and other lumber, he and a friend built a hide (also known as a bird blind) 20 feet off the ground and just 8 feet from the nest. They worked for a week, 45 minutes at a time, so the owls would grow accustomed to it.

He then waited a couple more days to ensure that owls were familiar with their new scenery. Alas, the new hide was near a field where the local townsfolk had gathered for a celebration, and several of them wandered over to the hide to see what was up. Hosking writes, tellingly, "Although none of them appreciated its true significance, they succeeded in thoroughly upsetting the owls, one of them whacking a wing across the back of a boy's neck."

Hosking used the hide the following night without incident, until he returned to check on his photographic equipment. "There was not a sound, not even a whisper of a wing," Hosking wrote. "But out of the silent darkness a swift and heavy blow struck my face. . . . The owl, with its night vision, had dive-bombed with deadly accuracy, sinking a deep claw into the center of my eye."

Hosking lost vision in that eye permanently.

When he returned to the hide several days later, the young owls had flown. At least, when the local authorities wanted to shoot the entire family of owls, he begged them not to. "The only shooting I wanted was with a camera, and the following year I did shoot them—on film. That, I thought, was just—an eye for a bird."

It's not fair to judge Hosking by today's standards when cameras and telephoto advances in technology have obviated the need to build a bird blind 8 feet from an owl's nest. That said, the fact that one of the owls had already attacked a boy should have been fair warning to Hosking to stay away.

The moral of all these stories is to stay away from nesting owls, especially around fledging time. If a screech owl dive-bombs you, get away and don't return for at least a week.

Keeping a Nature Journal

Hosting a screech owl family is rewarding, but if you want to increase your understanding and appreciation of what's going on, we recommend that you keep a nature journal. If possible, start before a screech magically peeks its head out of your nest box, and make it a habit. That way, if you get an owl, you'll have the right mindset to record on paper what you see and think.

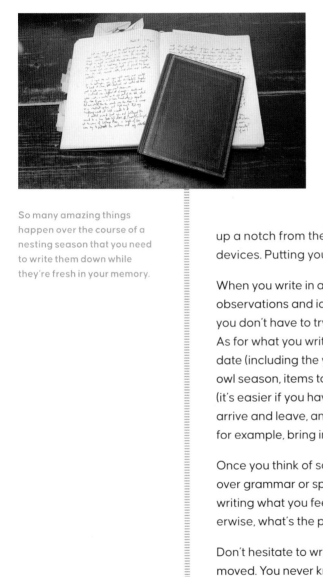

So many amazing things happen over the course of a nesting season that you need to write them down while they're fresh in your memory.

Don't skimp on the journal or your writing instrument. You don't have to invest in a large leather-bound journal. A high-quality 7 × 9 inch journal will get the job done. Throw in a rollerball pen with ink that doesn't clot or smear, and you're ready to write.

Why spend the extra money, you ask? Because you need to be in the right frame of mind to write, and you need to write thoughts in one place, not on scraps of paper or a legal pad. The journal and pen set the right tone. There's something about putting ink on paper—it takes the writing up a notch from the emails and texts we compose on our electronic devices. Putting your thoughts on paper makes you focus more.

When you write in a journal, jot things down right away, when your observations and ideas are fresh and often more insightful. That way, you don't have to try to remember what you saw and then write it down. As for what you write, spend a little time on the details: time of day, date (including the year), weather, even your mood if you like. During owl season, items to keep track of are which owl is roosting in the box (it's easier if you have a red-morph and a gray-morph one), when they arrive and leave, and the prey items they collect and when—they may, for example, bring in more items before a storm.

Once you think of something to write about, write quickly. Don't fret over grammar or spelling or those other hobgoblins that keep you from writing what you feel. Just make sure what you scribble is legible. Otherwise, what's the point?

Don't hesitate to write about other nature topics when you are so moved. You never know when they'll prove useful.

Tips from Screech-Box Hosts

Rather than just tell you what we've learned about hosting screech B&Bs and installing livestreaming cameras, we also asked several experienced nest-box hosts what they've learned. Their responses, edited for space, reflect the complexity of their setups.

A red-morph and a gray-morph Eastern Screech Owl at a nestbox opening.

Lisa Potash of Oakland, New Jersey, has had two owl boxes since 2015 and two livestreaming Green-Backyard nestcams since 2021. One is livestreamed on YouTube as "Owl in the Box." The female nests in an owl box in Lisa's front yard, while the male roosts nearby and caches prey items in a backyard nest box.

Lisa Potash has two owl boxes with inside cameras that are mounted on poles with baffles.

Mike Schouten's owl box has cameras inside and out.

Andrea and Michael Bean's trail camera captured this shot of a screech owl visiting the nest box.

Mike Schouten of Centerville, Utah, has had a screech box since 2014. He added a nestcam in 2018, and the first confirmed clutch arrived the following spring. He currently uses an Amcrest 1080P PTZ (pan-tilt-zoom) robotic video camera with 3× optical zoom, an Ethernet connection, and a Wi-Fi connection. He also has an Amcrest 4MP Outdoor PTZ (POE + IP camera, optical 25×, motorized) for his four-panel YouTube livestream. He installed a network switch on the nest tree, and all the cameras connect to that. You can learn more about Mike's setup and watch videos at bellanestcam.com.

Bart H. of central Florida has had nesting owls for 9 out of 10 years. He reports, "One year the female abandoned the nest with three eggs early on. It seemed something may have happened to the male that year." Bart says his original owl camera was "pretty crude, now long gone." He now uses a POE IP camera setup from Green-Backyard.com.

Andrea and Michael Bean of Peabody, Massachusetts, installed their owl box and Browning Dark Ops trail camera in 2018. The camera is mounted on an angled tree limb within 10 feet of the box, facing the entrance hole, and can be accessed with a small footstool. They put an SD card into the camera every day before dusk, record 10-second videos, and remove it in the morning after the owl(s) have returned to the box, usually around 7:00 a.m. After Andrea and Michael watch the videos and save the ones they want, they clear off the SD card and start the process again.

Laura and Ty Mayo of Grapevine, Texas, have had a screech box for more than 14 years. They added a nestcam after the first year and later replaced it with a Hawk-eye HD from Birdhouse Spy Cam.

Bill Powers of Murrysville, Pennsylvania, started the Eastern Screech Owl Research Project in 2011 to study the species' behavior in a natural environment during winter roosting and spring nesting. The project has eight screech owl boxes fitted with small cameras and one external pan-tilt-zoom camera used to watch the screeches hunt and enter and exit the owl boxes. The project has cameras in all eight boxes all year long to monitor the activity, and the videos can be viewed at pixcams.com/screech-owls. Some boxes have more than one camera.

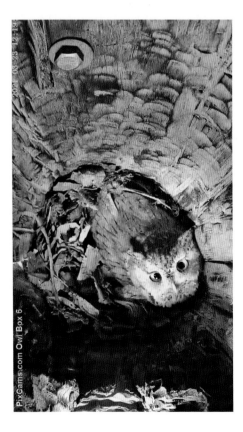

Above: This Eastern Screech Owl Research Project nest box features an addition to accommodate a side-view video camera.

Left: An owl sits inside one of the Eastern Screech Owl Research Project's eight nest boxes.

Would you recommend having a nestcam?

Lisa: If you're willing to put in the time and effort, yes. It opens the world of screech owls like never before. We started with one nest box. Then after having consistent owl pairs, we decided a second box would give them another option for nesting and winter roosting. We finally saw our first owl family fledge (without a nestcam), and that was the push I needed to install our two cameras.

This nesting season we were able to observe courtship, vocalizations, egg-laying and hatching, and some surprises, with a third owl visiting the pair in the box during brooding. Amazing stuff! Just be prepared for seeing things that can be upsetting. For example, watching little hatchlings succumb to cold and perish, with just one owl parent trying her best to feed them solo.

The view from Jim's nestcam.

Andrea: Yes, absolutely. We recommend a no-glow infra-red trail camera. It took a little bit to find the right location through trial and error. Once we did, we had no further issues. If you can put a nest box close to your home, an internal nestcam would be highly entertaining.

Laura: A camera isn't necessary if you just want to provide safe nesting habitat, but we wanted to observe as much as we could. We also wanted to stream it on our television so we could share it with our granddaughters and immerse ourselves in "everything screech owl." It has provided a wonderful insight into the miracle of nature.

Mike: I'd certainly recommend having a nest box and camera(s). The screech owls really don't seem to be bothered in any way by having cameras in the box or outside. Of course, the nest box should be a little larger to account for the cameras. My nest box is larger than most, with a perch inside that the owls love—especially when the eggs have hatched. They can be with the owlets to keep an eye on them.

Mike Schouten's livestream features multiple nest views on the same screen.

What have you learned?

Lisa: Screech owl pairs exhibit multilayered patterns of behavior with one another. They are such good mates and parents!

Bart: There are significant differences in how the owls operate in different areas of the country since their range is so wide. In Florida, I've found you need to put the boxes where they get shade during the heat of the day. April and May can be quite hot and dry. Nothing worse than watching owlets pant for days on end like they are going to die any minute.

Andrea: We learned that if successful in nesting, the owls will return each season. Roosting and nesting holes are extremely hard to come by, and they

This Eastern Screech Owl female in Grapevine, Texas, had an unprecedented 2021 season. Her first clutch failed due to unexpected freezing temperatures. Two weeks later, she laid a new clutch of four eggs—a total of eight eggs in 28 days. The second clutch wasn't viable, and the female didn't incubate. A week later, a third clutch was laid, with three owlets successfully fledged.

will defend the box with amazing tenacity. And we learned about their behavior and habits, along with their different vocalizations, and for each purpose that they're used.

Laura: Some harsh things happen in nature—survival of the fittest and all of that. When we started streaming our camera, we had five eggs hatch. One of them had a disease where its bone formation was severely compromised. We had to watch that little one struggle to survive, and of course it didn't. It was so painful, but I explained to my granddaughters, "Nature is harsh, but it's so beautiful. This is how things work, and not everything lives." It was a blessing to be able to use that as a life lesson.

Mike: Many cameras have a fixed focal distance. If the owl isn't within the focal range, it will tend to be out of focus. I was able to get around this by purchasing cameras that had zoom capabilities and autofocus.

What was your biggest surprise?

Lisa: When the owls roost together, how much allopreening (preening or grooming the other owl's feathers) goes on in the nest box.

Bart: The owls get used to humans being around and looking at them, but I never mess with them in the box except in an emergency. A few years ago when one owlet died and mom didn't remove it, bugs were taking over the remaining two owlets. I had to open the box and remove it while mom and dad were in a nearby tree. They weren't offended, but I don't recommend messing with them before the eggs hatch or they might quit.

Andrea: The biggest surprise was the amount and variety of food that the owlets consume. Earthworms make up at least 80 percent of the owlets' diet, followed by rodents, frogs. Very rarely are birds brought back to the box for the owlets. Also, we haven't had a single mouse in our attic during the winter, starting from the first season we had the nest box up.

Laura: I'll never forget one February. The owl laid her eggs, and we had a cold snap with temperatures well below freezing. And she kept leaving in the evening to go hunt. Of course, I was sitting in front of the television, lecturing her about how she should stay home and let the dad go hunt. Then we learned that the incubation period doesn't necessarily begin right away, and that mother knows best. That's a miracle also. How can those thin little eggs survive in 10-degree weather, and then hatch and be perfectly healthy?

Mike: Just nature itself. Hardly a day goes by when I don't wonder why the owls did this or that or acted a certain way, or how other birds really don't like owls.

What advice do you have for folks contemplating putting up an owl box?

Lisa: Figure out a way to keep out squirrels, have adequate ventilation, and make sure the roof seam is watertight.

Bart: Each yard is going to be different. I've found that screech owls require heavy cover nearby and a staging perch near the nest box to make it attractive. The male stays in a big oak tree about 12 feet from the box, and there is a perch about 6 feet in front of the box that they like to hunt from, watch over the box,

A parent has flown in from the nearby cover to feed an owlet at Bart's nest box.

and stage their trips back and forth. The longer open space that they have to cover, the less likely they will like the location.

Andrea: If you have the space, I'd recommend putting up two nest boxes. Screech owls are losing habitat quickly and need our help. If you can purchase or build a squirrel-resistant box, do it. If you can't, try to make it as squirrel-proof as possible. Give the owls their space and let them be as natural as possible—don't camp out under the box all day.

Mike: I recommend livestreaming the owls, or any birds for that matter. Millions of people love to watch birds online. It takes forever to build up viewers, but that's okay. Small audiences make the viewing more personal.

Anything you'd like to add?

Lisa: Your owl box's location isn't as important as making sure the owls have a clear pathway to and from the box. Putting the box in a spot with full sun all day isn't wise during nesting time but would be okay if the owls are using it for a winter roost.

Bart: Screech owls are certainly interesting and fun to watch and learn from. There is a responsibility that comes with the nest boxes. I can't put any on trees in Florida because the squirrels, snakes, and raccoons would get them or take over the box. The owl boxes are always under the gable end of the main house or aluminum outbuilding, where it's pretty much impossible for snakes and critters to get at them.

Mike: If possible, make the nest box easier to raise and lower [via pulleys, pivoting posts, etc.]. This will save you from falling from a ladder.

Andrea: If you decide to try a trail camera, be sure it is a true no-glow, with infra-red technology. Most important, enjoy the owls!

Laura: When we discovered we had an owl in the box, I thought this is a gift given to us by the universe, or God, or however you want to frame it. Now, all these years later, I think, "I hope we are gifted again in the spring. I hope we are allowed to participate in this divine event, and to continue to learn from it and share it with others."

APPENDIX:
Photographing Owls

The North American Nature Photography Association (NANPA), the American Birding Association (ABA), and the Nature Photographers Network (NPN) have online codes of ethics worth reading by photographers and bird-watchers alike. They are must-reads for anyone who cares about owls and other birds, nature, and our environment. Essentially, the codes boil down to "Have fun and take some great photos, so long as you do not adversely affect the birds or others' enjoyment of them. And protect the environment while you're at it." Who would argue with that?

Bird photography has come a long way since the 1830s, when Louis Daguerre coated a silver-plated piece of copper with a light-sensitive chemical and invented the first widely used photographic process.

The earliest bird photograph we've read about was of a Long-eared Owl, taken by an anonymous British photographer back in 1868. Owls have been favorite subjects ever since. That first owl picture was no small feat, given nineteenth-century technology. Early photographers used large cameras with photographic plates as large as 12 × 10 inches. The cameras were so cumbersome they were transported by wheelbarrow.

In 1903, ornithologist and writer Frank Chapman described his efforts to photograph an Eastern Screech: "In spite of the poor light prevailing at this hour, the bird's stillness has tempted repeated trials to secure its picture, and the most successful, made with a fourteen-inch lens and an exposure of fifteen seconds, is here shown."

The legendary artist Roger Tory Peterson photographed his first bird at age 13 or 14, using a costly Premo No. 9 camera that featured bellows and 4 × 5 inch plates. The subject? A screech owl that his friend had extracted from its roosting hole. In a move we guess would make the latter-day Peterson cringe, the boys tied a string around the owl's leg so it wouldn't fly away.

Now, for the cost of a 36-frame roll of Kodacolor, you can buy a reusable 32-gigabyte memory card that can be filled with hundreds or thousands of images over and over. You can take as many photos as you like, download them, and edit them within minutes. And if the bird turns out to be unusual, you can spread the news—with digital confirmation—lickety-split over the Internet.

The one downside to the digital photography boom is an uptick in the number of folks with cameras who ignore the basic rules of good birding. The cardinal rule, of course, is to stay away from an active nest. You may stress the birds and, in some cases, cause them to abandon the nest. Make every effort not to disturb an owl during its late winter to spring nesting season, a critical time in its yearly cycle.

Whether it's because owls are so photogenic or so seldomly seen by most people, many nature photographers can get crazed when they get a chance to photograph one, especially owls with their young in the nest. The dogged pursuit of a great owl photograph has become one of the most contentious subjects in the birding world—whether it's photographers trampling sensitive areas to get a closer shot or using store-bought rodents to get a staged photo of an owl pouncing on prey.

If you are fortunate enough to have a screech nesting on your property, you may have another concern. If word gets out that there's an opportunity to photograph a screech (or any owl), some people will do almost anything to get a photo. That includes trespassing on your property and harassing the owl.

Owl roosts have been blocked off, and conservation areas have had to close to the public at times to protect the owls. Many online birding forums and pages do not allow the posting of owl sightings. Although birders are known for their willingness to share bird sightings, many treat owl locations the way fishermen treat their favorite spots.

When aiming to observe and photograph screech owls, do so cautiously and respectfully. Many books and articles describe how to view owls, but too few talk about the importance of not flushing them. We like the advice offered by the folks at Project SNOWStorm, a national nonprofit group that tracks Snowy Owl peregrinations every winter: "Just because the owl may tolerate a fairly close approach doesn't mean you should push the envelope. If the owl is visibly reacting to your presence—fidgeting, repeatedly staring at you, head-bobbing, or changing position—you're too close and need to back off immediately." In short, err on the side of caution.

If you really want to see an owl up close or photograph one, we suggest checking with your nearest bird rehab center. Some places hold fundraisers where you can photograph their resident owls up close. Many of the birds cannot be released into the wild because of injuries, but they provide other services, from serving as ambassadors at live raptor shows to helping raise younger raptors that have been separated from their parents.

Visual encounters with owls in the wild are relatively rare, because they're camouflaged and they spend most of the day roosting high in trees, inside tree cavities, or in owl boxes. Owls can sometimes be viewed when crows, jays, chickadees, or other birds discover them in their territory. The birds will defend their domain by diving and calling repeatedly at the perched or flying owl, an activity called mobbing. Look and listen for this behavior, then look for the owl.

If a screech owl camouflages itself by stretching tall, with its ear-like tufts up and its wings pressed close to its back, it's employing a defense posture. It's pretending to be a dead stub on a tree branch, and you're too close. Similarly, if the owl flies away, then you have spooked it—and failed to be mindful of the owl. Please don't repeat your error. It's all about the owl, not a pretty picture or a better view. And, remember, a sleepy owl is a happy owl. Please don't snap a twig or squeak like a mouse to get the owl to open its eyes.

Nature writer and photographer Wayne Lynch uses a bird blind to get closer to roosting owls. "Of all the birds, owls are the ones you can watch fairly safely," Lynch says. "The problem is when the photographer isn't paying attention to what the owl is telling them, when the owl is stressed, and they don't move back. Blinds are the secret for me because the owl just keeps going about his business. You see intimate behavior you'd never see otherwise."

Many raptor rehab centers have fundraisers where photographers can take great close-ups of screech owls that would be impossible in the wild.

Opposite: Nature writer Wayne Lynch photographed these courting screech owls.

The farther away you are from a roosting owl, the better. This is one of the few circumstances when the photographic results from digiscoping—in which a digital camera is used to record distant images through the eyepiece of a spotting scope—can greatly exceed the results from a digital camera, even with a 400-mm lens. Why? Because there's no good way to get close to your subject without stressing it out. If your photographic setup doesn't have enough magnification to satisfy you, don't take it out on the bird. Encroaching shouldn't be an option. Look for a more conducive roost somewhere else.

Fortunately, you can get a quality photo without breaking the bank. If you have a spotting scope or binoculars, you can buy adapters for cellphones and other cameras to take surprisingly good digiscope or digibin images. Some scopes and binoculars also have image-capturing capabilities.

Five Tips for Great Owl Photos

1. Know your owls. The more you know, the better your photo will be. Sometimes photos of an owl may be off even though they're in focus, well lit, and well composed for the simple reason that the owl is in an odd pose or is stressed. If the owl looks bug-eyed, you were likely stressing it out, and the photo is not nearly as attractive to an experienced eye.

Using a flash to take photos of this roosting screech caused it to abandon the owl box.

2. Avoid using a flash. The light often looks artificial, the eyes often have an unnatural look, and it can annoy the bird. In at least one instance, at the beginning of nesting season thus use of a flash likely caused a screech owl to relocate. Try cranking up the ISO on your digital camera instead. Or, if you're incredibly persistent or extraordinarily motivated, use a tripod and exposures of 1 second or even longer. Yes, owls can stay still for that long, a testament to their stealth.

3. Go easy on the Photoshop. This is a personal preference—perhaps it's because we spend too much time in front of our computers as it is. Software to enhance/improve digital images is great, but keep it minimal and subtle.

4. Think in black-and-white sometimes. Want owl photos unlike most you see? You can start by converting some of your best color images of screech owls to black-and-white. All it takes is a click on your laptop screen. You'll see the image in a whole new light.

5. Don't overdo it. Once you get the shot you want, stop clicking the shutter and quietly depart. You may be disturbing the owl, and you'll only have to delete most of the images later. Also, unless you're a terrific photographer and people show a genuine interest, there's no need to show anyone more than a few of your owl shots. Life's too short.

Acknowledgments

To my wife, Patty, for her insights, encouragement, and patience.

Special thanks to the MFP, Kevin Watson, Stiles Thomas, Joan Brookbank, Stacee Lawrence, Lisa Potash, Rick Radis, Ken Wiegand, Chris Soucy and The Raptor Trust, Tony Angell, Robyn Bailey, Andrea Bean, Jim Belthoff, Lisa Brousseau, Joseph Bruchac, Matt Cuda, Michael Gardner, Bart H., Cliff Hagen, Linnea Hall, Daniel Klem, Ted Kooser, Nic Korte, Jacoba Lawson, Pat Leonard, Deane Lewis, Wayne Lynch, Laura Mayo, Kevin McLain, Dorcas MacClintock, Andrew McGann, Sarah Milhollin, Lisa Owens-Viani, Tina Phillips, Robert Peck, Bill Powers, Annick Gionet Rollick, Mike Schouten, Steven Sullivan, Jason Weckstein, Jenohn Wrieden, Paul Yorke, Jim Zipp, and the Lee Memorial Library in Allendale.

I would also like to thank Hawk Mountain Sanctuary's Acopian Center, where I was writer-in-residence in March 2022.

—Jim Wright

A huge thank you to my wife, Jennifer, and our four kids, who put up with my quirkiness and for all the owl walks we enjoyed together. Special thanks to (late) John and Sarah Cushing, Debby Ramsey, Andrea Bean, Jason and Emily Reulet, Sonia Silk, Georgeanne Richards, Katherine Davis, Steve Grinley and Margo Goetschkes, Jim Berry, Phil Brown, Cliff Shackelford, Kevin and Linda Greeley, Jason and Heather Wright, and G.V. and Beth Weston.

–Scott Weston

The two range maps use data from the eBird Status and Trends Project at the Cornell Lab of Ornithology, eBird.org. Any opinions, findings, and conclusions or recommendations expressed in this material are those of the author(s) and do not necessarily reflect the views of the Cornell Lab of Ornithology.

Fink, D., T. Auer, A. Johnston, M. Strimas-Mackey, O. Robinson, S. Ligocki, W. Hochachka, L. Jaromczyk, C. Wood, I. Davies, M. Iliff, and L. Seitz. 2021. eBird Status and Trends, Data Version: 2020; Released 2021. Cornell Lab of Ornithology, Ithaca, New York. Available via doi. org/10.2173/ebirdst.2020

Chapter 3 ("The Magic of Owls"): passages about feathers were adapted from Jim Wright's "Bird Watcher" column, published in *The Record* (North Jersey) on February 4, 2021.

References

Books

Ackerman, Jennifer. 2017. *The Genius of Birds*. Penguin Press.

Adler, Bill, Jr. 1996. *Outwitting Squirrels*. Chicago Review Press.

Angell, Tony. 1974. *Owls*. University of Washington Press.

Angell, Tony. 2015. *The House of Owls*. Yale University Press.

Audubon, John James. 1832. *Birds of America*. Havell.

Baird, Spencer Fullerton, et al. 1875. *A History of North American Birds*. Little Brown & Co.

Bent, Arthur Cleveland. 1938. *Life Histories of North American Birds of Prey. Part 2*. U.S. Government Printing Office.

Blanchan, Neltje. 1898. *Birds That Hunt and Are Hunted*. Doubleday & McClure Co.

Buechler, James. 1962. *The Clue of the Screeching Owl*. Grosset & Dunlap.

Burgess, Thornton. 1919. *The Burgess Bird Book for Children*. Little, Brown & Co.

Burroughs, John. 1871. *Wake Robin*. Hurd & Houghton.

Cameron, Angus, and Peter Parnall. 1971. *The Nightwatchers*. Four Winds Press.

Catesby, Mark. 1731–1743. *The Natural History of Carolina, Florida, and the Bahama Islands*. Vol. 1.

Chapman, Frank M. 1895. *Handbook of Birds of Eastern North America*. D. Appleton & Co.

Condon, Marlene A. 2006. *The Nature-Friendly Garden*. Stackpole Books.

Dunne, Pete. 2006. *The Art of Pishing*. Stackpole Books.

Gehlbach, Frederick R. 1994. *The Eastern Screech Owl*. Texas A&M Press.

Gill, Frank B. 1994. *Ornithology*. Worth Publishers.

Gladden, George. 1917. *Birds of America*. University Society.

Harrison, George H. 2000. *Squirrel Wars*. Willow Creek Press.

Heinrich, Bernd. 2016. *One Wild Bird at a Time*. Houghton Mifflin Harcourt.

Henshaw, Henry W. 1914. *Common Birds of Town and Country*. National Geographic Society.

Hosking, Eric. 1970. *An Eye for a Bird*. Hutchinson & Co.

Johnsgard, Paul A. 1988. *North American Owls*. Smithsonian Institution Press.

Klem, Daniel. 2021. *Solid Air*. Hancock House.

Kooser, Ted. 2004. *Delights and Shadows*. Copper Canyon Press.

Kricher, John. 2020. *Peterson Bird Behavior*. Houghton Mifflin Harcourt.

Lynch, Wayne. 2007. *Owls of the United States and Canada*. John Hopkins University Press.

Maslow, Jonathan. 1983. *The Owl Papers*. E.P. Dutton.

McClung, Robert. 1959. *Otus*. William Morrow & Co.

Pearson, Gilbert, ed. 1936. *Birds of America*. Garden City Publishing Co.

Peterson, Roger Tory. 2006. *All Things Reconsidered*. Houghton Mifflin.

Pipelow, Nathan. 2017. *Peterson Field Guide to Bird Sounds of Eastern North America*. Houghton Mifflin Harcourt.

Sibley, David Allen. 2020. *What It's Like to Be a Bird*. Alfred A. Knopf.

Tate, Peter. 2007. *Flights of Fancy*. Random House Books.

Thompson, Ernest E. 1891. *Birds of Manitoba*. U.S. Government Printing Office.

Thoreau, Henry David. 1854. *Walden*. Ticknor and Fields.

Thorington, Richard W., Jr., and Katie Ferrell. 2006. *Squirrels*. Johns Hopkins University Press.

Unwin, Mike. 2017. *The Enigma of the Owl*. Yale University Press.

Webb, Walter F. 1895. *Ornithologists' and Oologists' Manual*. Albion, N.Y.

Weidensaul, Scott. 2015. *Peterson Reference Guide to Owls*. Houghton Mifflin Harcourt.

Whittaker, Diane. 2022. *The Secret Perfume of Birds*. Johns Hopkins University Press.

Wright, Mabel Osgood. 1895. *Birdcraft*. The Macmillan Company.

Newspaper, Magazine, and Journal Articles

Allen, M. L., and A.P. Taylor. 2013. First record of scavenging by a Western Screech Owl. *Wilson Journal of Ornithology*, 125(2).

Altoona Times. 1885. The bounty for wild animals, etc. (Altoona, Pennsylvania), Nov. 3, 1885.

Audubon, John J. **1827–1838**. Little Owl (Saw-Whet), John J. Audubon's Birds of America. Audubon.org, Plate 199.

Beebe, F.L. 1947. Mystery birds of dusk and darkness. *The Vancouver Sun*, Jan. 18.

Belthoff, James R., and Gary Richison. 1990. Roosting behavior of postfledging Eastern Screech Owls. *The Auk*, 107(3).

Belthoff, James R., and Gary Richison. 1989. Natal dispersion of Eastern Screech Owls. *The Condor*, 91(2).

Belthoff, James R., and Gary Richison. 1990. Nest-site selection by Eastern Screech-Owls in central Kentucky. *The Condor*, 92(4).

Belthoff, James R., Earl J. Sparks, and Gary Ritchison. 1993. Home ranges of adult and juvenile Eastern Screech-Owls: size, seasonal variation and extent of overlap. *Journal of Raptor Research*, 27(1).

The Commonwealth. 1905. Photograph themselves. (Greenwood, Mississippi), July 21.

Ellsworth, E., and James R. Belthoff. 1999. Effects of social status on the dispersal behaviour of juvenile Western Screech-Owls. *Animal Behaviour*, 57.

The Evening Review. 1916. Attacked by an owl. (East Liverpool, Ohio), July 14.

Fisher, Harold. 2016. Northern Hawk Owls use nest box. *Blue Jay Journal*, 74(4).

Forbush, Edward Howe. 1905. Nesting-boxes. *Bird-Lore*, Jan./Feb.

Forbush, Edward Howe. 1927. As quoted in. Bent's *Life Histories of North American Birds of Prey. Part 2*. U.S. Government Printing Office, 1938.

Gehlbach, Frederick R. 1994. Nest-box versus natural-cavity nests of the Eastern Screech-Owl. *Journal of Raptor Research*, 28(3).

Gosselin, Michel, and Kristen Keyes. 2009. A Long-eared Owl × Short-eared Owl specimen from Ontario. *Ontario Birds*, 27(1).

Hahn, D. Caldwell. 2011. Patterns of maternal yolk hormones in Eastern Screech Owl eggs. *General and Comparative Endocrinology*, 172(3).

Hardy, Paul C., and Michael L. Morrison. 2003. Nest-site selection by Western Screech Owls in the Sonoran Desert, Arizona. *Western North American Naturalist*, 63(4).

Harrop, Froma. 2016. Rethink the naming of wild animals. *The Denver Post*, July 16.

Hasbrouck, E.M. 1893. Evolution and dichromatism in the genus *Megascops*. *The American Naturalist*, 27(318).

Herting, Brian L., and James R. Belthoff. 2001. Bounce and double trill songs of male and female Western Screech-Owls: characterization and usefulness for classification of sex. *The Auk*, 118(4).

Johnson, David H., et al. 2010. Users guide to installation of artificial burrows for burrowing owls. Tree Top Inc. Available via wdfw.wa.gov/sites/default/files/publications/01199/wdfw01199.pdf.

The Kansas City Star. 1903. A bird takes pictures. (Kansas City, Kansas), Dec. 30.

Klatt, Paul H., and Gary Ritchison. 1993. The duetting behavior of Eastern Screech Owls. *The Wilson Bulletin*, 105(3).

Kwok, Roberta. 2017. The "smart nest box" provides a peek into the secret lives of birds. *Audubon Magazine*, Summer.

Larson, Christina. 2022. Rare Snowy Owl soars over Washington, thrills crowds. *The Associated Press*, January 10.

Lebanon Daily News. 1929. Screech owls attack people. (Lebanon, Pennsylvania), Aug. 6.

Long, Linda L., and Jared D. Wolfe. 2018. A briefing paper on the science directly related to the effects of Barred Owls on Spotted Owls. USDA Forest Service, Redwood Sciences Laboratory. https://www.fs.usda.gov/treesearch/pubs/58777

Marini, Miguel Ângelo, Linnea Hall, et al. 2020. The 5 million bird eggs in the world's museum collections are an invaluable and underused resource. *The Auk*, 137(4).

Marshall, Joe T. 1967. Parallel variation in North and Middle American screech owls. *Monographs of the Western Foundation of Vertebrate Zoology*, No. 1, July.

McCallum, D. Archibald, Frederick R. Gehlbach, and Sophie W. Webb. 1995. Life history and ecology of Flammulated Owls in a marginal New Mexico population. *Wilson Bulletin*, 107(3).

McKinney, Bonnie. 2008. Black Gap Elves: My three-year adventure with the smallest owl in North America. *Texas Parks and Wildlife Magazine*, May.

Oleyar, Dave. 2017. Following forest owl community and population dynamics in different forest types in the western United States. *Earthwatch 2017 Annual Field Report*. Hawk Watch International. https://earthwatch.org/sites/default/files/2020-03/earthwatch-field-following-forest-owls-western-us-2017.pdf

Pinson, Jerald. 2019. Purple martins, Native Americans formed lasting bonds. *Austin Statesman*, July 24.

Proudfoot, Glenn A., Sam L. Beasom, and Felipe Chavez-Ramirez. 1999. Biology of Ferruginous Pygmy-Owls in Texas and application of artificial nest structures. Texas A&M University-Kingsville, *Wildlife Management Bulletin*, No. 5.

Roberts, Hilda. 1927. Louisiana superstitions. *The Journal of American Folklore*, 40(156).

Roberts, Thomas S. 1899. The camera as an aid in the study of birds. *Bird Lore*, 1(1), February.

Rogue Valley Audubon Society 2017. Great Gray Owl nest platform project. Rogue Valley Audubon Society. https://www.roguevalleyaudubon.org/great-gray-owl-project/

Salt Lake Telegram, 1934. It's th' Hoot. (Salt Lake City, Utah), June 28.

Smith, Dwight G., and Raymond Gilbert.. 1984. Eastern Screech-Owl home range and use of suburban habitats in southern Connecticut. *Journal of Field Ornithology*, 55(3).

South Pasadena Foothill Review. 1930. Telephone repairman injured in battle with owl. (South Pasadena, California), May 2.

Sparks, Earl J., James R. Belthoff, and Gary Ritchison. 1994. Habitat use by Eastern Screech-Owls in central Kentucky. *Journal of Field Ornithology*, 65(1).

Taverner, P.A. 1923. Bird houses and their occupants. *Calgary Herald*, Dec. 29.

Washington Post. 1914. New Home in Washington Park. (Washington, DC), March 29.

Wilson, Eddie W. 1950. The owl and the American Indian. *The Journal of American Folklore*, 63(249).

Videos

BBC. *The World's Largest Long-Eared Owl Roost*, short documentary, December 2018.

Kristersson, Mikael. *Kestrel's Eye*, documentary, 1999.

OwlCam, The Hidden World, documentary, 2005.

Rober, Mark. *Backyard Squirrel Maze 2.0: The Walnut Heist*. YouTube Video.

Photography Credits

The three principal photographers for The Screech Owl Companion are Kevin Watson, Scott Weston, and Jim Wright.

Kevin's photos are on pages 38, 41 (bottom right), 44 (2), 47 (top), 52, 56 (bottom), 61, 75, 76, 79, 81, 83, 85, 87, 88, 90, 91, 93, 94, 96, 98, 99, 100, 109, 124 (top), 199, 202, and 228.

Scott's photos are on pages 13, 14, 27, 36, 37, 59, 68 (bottom), 106, 110, 114 (4), 122, 125 (2), 130, 132, 135, 138 (4), 142, 144, 150, 155, 159, 160, 166, 181, 182, 183, 188, and 204.

Jim's photos are on pages 41 (left), 47 (bottom), 51 (right), 54 (right), 68 (top and middle), 74, 104, 105 (left), 112, 119, 126, 140, 148 (2), 154, 157, 162, 163, 164, 171 (top), 173, 187, and 193.

Other images are credited to:

All Canada Photos / Alamy
Stock Photo, page 99

Angell, Tony, from his book Owls, page 102

Audubon, John James, Biodiversity
Heritage Library, Wikimedia
Commons, page 26 (bottom left)

Barrack, Jerry, pages 2, 8, 179 (4), 183

Bean, Andrea, page 191 (bottom)

Bird-Lore Magazine, 1905, page 128 (top)

Blanchan, Neltje, Birds that Hunt
and Are Hunted, page 54 (left)

Carnegie Museum of Natural
History, Powdermill Avian
Research Center, page 43

Catesby, Mark, The Natural History of
Carolina, Florida and Bahamas V.
1, Wikimedia Commons [File:The-
little-owl.jpg], page 26 (top left)

The Commonwealth, Greenwood,
Miss., page 152

Cornell Lab of Ornithology, Data
from eBird Status and Trends (see
Acknowledgments), page 21

Cornell Lab of Ornithology, Macaulay
Library, vocalization by an Eastern
Screech owl, sound recording
by Wil Hershberger, page 34

Cornell Lab of Ornithology, Macaulay
Library, vocalization by a Western

Screech owl. sound recording
by Bruce Rideout, page 34.

Cornell Lab of Ornithology,
NestWatch, page 58 (2),

Cuda, Matt, page 33

Cushing, John D., pages 29
(right), 62 (right)

Davis, Katherine C., page 180

Deviche, Pierre, page 24

Decatur Herald, 1928, page 128 (bottom)

Dixon, Franklin W., Hardy Boys:
Case of the Screeching Owl,
Grosset & Dunlap, page 20

The Evening Review (East Liverpool,
Ohio), "Attacked an Owl," page

Finley, William L., Photographs
Collection, circa 1900-1940 (Org.
Lot 369, OHS), page 200

Fuertes, Louis Agassiz, Wikimedia
Commons [File:FMIB 43290
Screech Owl- Upper figure, gray
phase; lower figure, rufous phase
of plumage.jpeg], page 170

Gehlbach, Fred, cover, The
Eastern Screech Owl, Texas
A&M Press, page 17.

Gardner, Steven Michael, GardnerGallery.
com, pages 72, 228

The Owl cover, Gold Key, Western
Printing Co., page 40

U.S. Department of the Interior, public domain, page 184

The Feather Atlas of North American Birds, U.S. Fish & Wildlife Service, page 23

Valette, Claude, Wikimedia Commons [File:26 HibouGravureAuDoigt.jpg], page 41 (top right)

Vaughn, Greg / Alamy Stock Photo, page 98

Washington Post, "New Home in Washington Park," March 29, 1914, page 115

Weston, Jenn, page 228

Wilson, Alexander, Biodiversity Heritage Library, Wikimedia Commons, [File:American ornithology; or, The natural history of the birds of the United States (1870) (17971914968).jpg] page 26 (right)

Wrieden, Jenohn Michele, page 22 (bottom right)

Yorke, Paul, page 146

Zipp, Jim (owl coughing up pellet), page 51 (left)

Note: As we mentioned in the Appendix, photographing owls up close in the wild can be very stressful for them. To avoid this as best as possible, the photos on the front cover, pages 20, 24, 27, 88, 199, and 202 feature owls from The Raptor Trust, an avian rehabilitation facility in Millington, NJ, and similar facilities, under the supervision of expert biologists and naturalists.

• • • • • • • • • • • • • • • • • • •

Index

Gmelin, Johann Friedrich, 25
gourds as birdhouses, 128
Grand Junction, Colorado, 28
grass owl, 84
Great Grey Owl, 92–93
Great Horned Owl
    nesting baskets, 103
    pellets, 50
    poisoning of, 68
    population, 28, 74
    sense of smell, 48
    species profile, 74–75
    tufts ("horns"), 44
Green-Backyard.com, 157
ground nesting, 85

H

H., Bart (screech-box host), 191,
    196–197
habitat, 28–29, 31, 61–63. See also
    species profiles
Hagen, Cliff, 70
Hahn, D. Caldwell, 174
Handbook of Birds of Eastern
    North America (Chapman), 48
Hardy Boys book, 20
Harrop, Froma, 177
*Harry Potter and the Half-Blood
    Prince*, 16
*Harry Potter and the Order of the
    Phoenix*, 16
*Harry Potter and the Philosopher's
    Stone*, 16
hat decorations, 54
HDOnTap, 155
height
    Eastern and Western Screech
      Owls, 18
    North America's tallest owl, 92
    ranges, 40
herbicides, 67, 69

A History of North American Birds
    (Baird, Brewer, and Ridgway),
    25
hives, 120
hobgoblin owl, 89
Hokkaido Island (Japan), 60
hoot owl, 74, 77
hornets, 120
Hosking, Eric, 185–186
The House of Owls (Angell), 49
howdy bird, 82
Hudsonian owl, 90
hunting of owls, 53, 55
hydration, 30

I

iguanas, 120
injuries
    avian foster families, 66
    contacting wildlife rehabilitators,
      66–67, 149
    displaced nestlings, 66
    picking up and transporting an
      injured owl, 67
    poisons, 68–69
    raptor rehab centers, 62, 64,
      65–66, 201–202
    treating stunned birds, 65–66
    vehicle impacts, 63
    windowpane impacts, 64–66
insecticides, 68–69, 121
insectivores, 95, 99
insect pests, 120–121
insects, stinging, 120
intelligence, 49

J

Jackdaw, 128
Johnson, Chris, 153
Jones, Lynd, 169

journal keeping, 186
The Journal of American Folklore,
    56

K

Kay Electric Co., 36
Keepers of the Earth (Bruchac), 55
Kennicott, Robert, 19
kestrel boxes, 103, 105
kestrels, 118
Kestrel's Eye (film), 152, 153
Kikinda, Serbia, 86
Klem, Daniel, 64
Kristersson, Mikael, 152
Kyle, Georgean, 63, 64
Kyle, Paul, 63, 64

L

ladder safety, 143, 197
Lapland owl, 92
leucism, 22
Levad, Rich, 28
Lewis, Deane, 44
life expectancy, 18
*Life Histories of Familiar North
    American Birds* (Bent), 169
Linnaeus, Carl, 25
little dukelet, 19
little horned owl, 19
little nightbird, 78
little owl, 78
"Little Owl Who Lives in the
    Orchard" (Oliver), 184
lizards, 120
Long-eared Owl, 19, 39, 86–87, 199
Lynch, Wayne, 49, 202

M

Marshall, Joe T., Jr., 42
marsh owl, 84
mating / breeding. *See also* nesting
    season
  breeding season, 173
  description, 169
  hybridization, 77, 84
  for life, 22, 169
  mating call, 32–33
  reproductive age, 18, 184
Mayo, Laura and Ty (screech-box
    hosts), 192, 193–194, 195, 196
McClung, Robert M., 176, 177
Megascops, 20
Megascops asio, 19, 21
Megascops kennicottii, 19, 21
Megascops petersoni, 22
Megascops stangiae, 22
Middle American Screech Owl, 20
Migratory Bird Treaty Act of 1918,
    45, 55
mites, 120
Molly and McGee livestream, 89
Mottled Owl, 25
mountain pygmy owl, 91
mouse owl, 19

N

names
  "Ace", short for *Asio*, 177
  bird banders' initials, 19
  "Kenny", short for *kennicottii*, 177
  misidentifications in literature,
    19–20
  Molly and McGee, 89
  naming wild animals, 177
  Ward and June, 152
names, common. *See* species pro-
    files

National Audubon Society
  on birdbaths, 62
  Bird-Lore articles on birdhouses,
    128–129
  Christmas Bird Count, 28, 70–71
  on helping the Eastern Screech
    Owl, 28
Native American peoples
  superstitions, 55–56
  Tlingit bowl featuring carved
    owls, 56
  use of hollow gourds, 128
*The Natural History of Carolina,
    Florida, and the Bahama
    Islands*, Vol. 1 (1731) (Catesby),
    23–24
nature, observations on ways of,
    195, 196
Nature-Friendly Garden (Condon),
    61
Nature Photographers Network
    (NPN), 198
nest boxes. *See* owl boxes / nest
    boxes
nestcams
  cellphone cameras and selfie
    sticks, 161
  choosing, 156–158
  citizen science use, 60
  Czech University of Life Sciences
    nestcam project, 153
  early remote techniques, 152
  evolution of, 152–154
  focal range, 195
  infra-red illuminated, 156
  installing, 126, 154, 161–165
  IP setups, 162, 191
  as lifesaving, 156
  limiting owl disturbance, 154–155
  livestreaming, 154–155, 156, 158,
    161–162, 165, 190, 197
  livestreaming multiple views, 194
  Molly and McGee livestream, 89

  placement in owl boxes, 105, 107
  power sources, 156
  privacy issues, 154–156
  screech-box hosts' experiences,
    190–197
  video storage, 161
  wildlife camera websites, 158
nestcam wiring
  2006 methods, 153–154
  2022 methods, 162–163
  Ethernet cables, 157, 162, 163,
    164–165
  and location of nestcam, 105
  surge protectors, 164–165
  Wi-Fi, 158, 159, 163, 191
nest collecting, 57
nesting season. *See also* mating /
    breeding
  courtship call notes, 35
  depiction in literature, 176
  male roosting behavior, 107, 158,
    169, 172, 175, 190
  staking out territory, 170
  timing of yard work, 62
nestlings
  calls and songs, 33
  dominance hierachy, 178
  survival rate, 60
  waste, 50
Nest-Record Card Program, 58, 60
nests
  cavity nests, 28, 31, 58, 80, 91,
    123–124
  former nests of other birds, 87,
    90, 95
  prospective sites, 102, 169
  underground, 82
Nest video camera, 158
NestWatch citizen science program,
    60, 103, 124, 153, 154
The Nidiologist, 57
nighthawk, 40
night hunting, 40, 46

evergreens and trees with dense foliage, 31, 61, 78–79, 80, 86, 89, 201
kestrel box, 105
in natural cavities, 31
roadside, 105
signs of, 182
Royal Society for the Protection of Birds, 128

S

saferodentcontrol.org, 68
safety, personal
  in owl box installation, 75, 142–143, 197
  when evicting squirrels, 114–115
Sandford, Lloyd, 176
Sand Martin, 128
Schouten, Mike (screech-box host), 191, 194, 195, 196, 197
scops owl, 20
screech-box hosts, 190–197
screech owl (in Europe), 89
screech owl species in North America, 20–21
The Secret Perfume of Birds (Whittaker), 48
SGARs (second-generation anticoagulant rodenticides), 68
Shackelford, Cliff, 124
Sharp-shinned Hawk, 20
shivering owl, 19, 56
Short-eared Owl, 84–85
shrews, 157
Sibley, David Allen, 42
sightings
  Barred Owl, 108
  cautions, 167
  eBird app, 102, 107
  experience of, 10–11, 70–71, 108–109, 125
  posting online, 60, 102, 107, 201

silver owl, 89
skunks, 118
smell, sense of, 48
snakes, 31, 116, 120, 176, 197
Snowy Owl, 95–97, 201
Solid Air (Klem), 64
Sona-Graph, 36
songs. *See* calls / songs
Soucy, Chris, 62, 66
South American screech owl, 22
sparrow owl, 78
species profiles, 73–99
Spotted × Barred Owls, 77
Spotted Owl, 97–98
squirrels
  American red squirrel, 112
  commandeering of owl boxes, 111, 112–115, 131–132
  eastern gray squirrel, 112, 113, 114
  flying squirrel, 112, 113, 114
  fox squirrel, 112, 113, 114
  gray squirrels, 113, 176
  infestation of Lafayette Square (Washington, DC), 115–116
  obstacle course video, 112
  Outwitting Squirrels, 112
  owl-squirrel conflicts, 113
  pine squirrel, 113
  preventative measures, 116–117
  preventative measures not recommended, 117–118
  as prey, 113
  red squirrels, 114, 176
  safety when evicting, 114–115
  shelters (dreys), 113
  as squatters, 63
  squirrel-resistant box owl boxes, 113, 131–139
  squirrel-resistant feeders, 116–117
  Squirrel Wars, 112
  types competing for owl housing, 112–113, 114
  western gray squirrel, 112, 113
Stang, Dorothy, 22

starlings, 118, 159
steeple owl, 89
Strigidae, 19
striped owl, 77, 92
suburban / urban areas, 83, 108
Sullivan, Steven, 117
Supercircuits, 153
superstitions about owls, 31, 55–56
swamp owl, 77
Systema Naturae (Linnaeus), 25

T

Tate, Peter, 31
Taverner, P. A., 128
Tawny Owl, 128, 185
territory
  of an owl pair, 107, 123, 169, 184
  behavior of specific owls, 32, 77, 99
  call when defending, 32–33
  dispersal season, 32, 182
  size, 28
  staking out, 170
Thomas, Adrian, 128
Thomas, Stiles, 108–109
Thompson, Ernest E., 30
Thoreau, Henry David, 12
tiger of the sky, 74
titmice, 49
Tlingit bowl, 56
Tlingits, superstitions, 55
Trail, Pepper, 45
tree cavities
  availability and owl use, 58, 102
  cavity-nesting birds, 118
  cavity nests, 28, 31, 58, 80, 91, 123–124
  emulating, 123
  old woodpecker cavities, 62, 80, 90, 91, 95, 99, 123, 124
  record of screeches using, 58, 60
Tree Swallow box, 153

The circles denote the eleven owls lurking in the
illustration by Steven Michael Gardner on Page 72.

Jim Wright is the author of several books about nature, a prize-winning journalist, and the birding columnist for *The Record*, North Jersey's leading newspaper. He has hosted screech owls in his backyard nest box since 2004 and observed them via a nestcam for more than a decade. He is leading a campaign to ban the rat poisons that kill owls and other wildlife. He lives in Allendale, New Jersey, where he is a marsh warden.

Scott Weston has been involved with birding and owling for over thirty-five years, going on predawn Audubon Christmas bird counts since he was eleven. In 2009, using squirrel-resistant nest boxes of his own design, he set up a screech owl trail that has fledged over 130 owlets. An accomplished photographer, Scott has contributed to local and national publications, including Cornell's *All About Backyard Birds*. He lives in northeastern Massachusetts.